Anonymous

A Treatise on Meteorological Instruments

Explanatory of their Scientific Principles, Method of Construction and Practical

Utility

Anonymous

A Treatise on Meteorological Instruments
Explanatory of their Scientific Principles, Method of Construction and Practical Utility

ISBN/EAN: 9783337338923

Printed in Europe, USA, Canada, Australia, Japan

Cover: Foto ©berggeist007 / pixelio.de

More available books at **www.hansebooks.com**

A TREATISE

ON

METEOROLOGICAL INSTRUMENTS:

EXPLANATORY OF

THEIR SCIENTIFIC PRINCIPLES,

METHOD OF CONSTRUCTION, AND PRACTICAL UTILITY.

BY

NEGRETTI & ZAMBRA,

METEOROLOGICAL INSTRUMENT MAKERS TO THE QUEEN, THE ROYAL OBSERVATORY, GREENWICH,

THE BRITISH METEOROLOGICAL SOCIETY, THE BRITISH AND FOREIGN GOVERNMENTS,

ETC. ETC. ETC.

LONDON:

PUBLISHED AND SOLD AT NEGRETTI & ZAMBRA'S ESTABLISHMENTS:

1 HATTON GARDEN, E.C., 59 CORNHILL, E.C., 122 REGENT STREET, W.,

AND 153 FLEET STREET, E.C.

1864.

Price Five Shillings.

PREFACE.

THE national utilisation of Meteorology in forewarning of storms, and the increasing employment of instruments as weather indicators, render a knowledge of their construction, principles, and practical uses necessary to every well-informed person. Impressed with the idea that we shall be supplying an existing want, and aiding materially the cause of Meteorological Science, in giving a plain description of the various instruments now in use, we have endeavoured, in the present volume, to condense such information as is generally required regarding the instruments used in Meteorology; the description of many of which could only be found in elaborate scientific works, and then only briefly touched upon. Every Meteorological Instrument now in use being fully described, with adequate directions for using, the uninitiated will be enabled to select those which seem to them best adapted to their requirements. With accounts of old or obsolete instruments we have avoided troubling the reader; on the other hand, we were unwilling to neglect those which, though of no great practical importance, are still deserving of notice from their being either novel or ingenious, or which, without being strictly scientific, are in great demand as simple weather-glasses and articles of trade.

We trust, therefore, that the work (however imperfect), bearing in mind the importance of the subject, will be acceptable to general readers, as well as to those for whose requirements it has been prepared.

The rapid progress made in the introduction of new apparatus of acknowledged superiority has rendered the publication of some description absolutely necessary. The Report of the Jurors for Class XIII. of the International Exhibition, 1862, on Meteorological Instruments, fully bears out our assertion, as shown by the following extract:—

"The progress in the English department has been very great;—in barometers, thermometers, anemometers, and in every class of instruments. At the close of the Exhibition of 1851, there seemed to have arisen a general anxiety among the majority of makers to pay every attention to all the essentials necessary for philosophical instruments, not only in their old forms, but also with the view of obtaining other and better forms. This desire has never ceased ; and no better idea can be given of the continued activity in these respects, than the number of patents taken out for improvements in meteorological instruments in the interval between the recent and preceding exhibitions, which amount to no less than forty-two."

*   *   *   "In addition to numerous improvements patented by Messrs. Negretti and Zambra, there is another of great importance, which they did not patent, viz. enamelling the tubes of thermometers, enabling the makers to use finer threads of mercury in the construction of all thermometers ; for the contrast between the opaque mercury and the enamel back of the tubes is so great, that the finest bore or thread of mercury, which at one time could not be seen without the greatest difficulty, is now seen with facility; and throughout the British and Foreign departments, the makers have availed themselves of this invention, the tubes of all being made with enamelled backs. It is to be hoped that the recent exhibition will give a fresh stimulus to the desire of improvement, and that the same rate of progress will be continued."

To fulfil the desire of the International Jury in the latter portion of the above extract will be the constant study of

NEGRETTI & ZAMBRA.

1st January, 1864.

TABLE OF CONTENTS.

CHAPTER VII.

SELF-REGISTERING THERMOMETERS.

CHAPTER VIII.

RADIATION THERMOMETERS.

CHAPTER IX.

DEEP-SEA THERMOMETERS.

CHAPTER X.

BOILING-POINT THERMOMETERS.

TABLES.

ADDENDA.

METEOROLOGICAL INSTRUMENTS.

In the pursuits and investigations of the science of Meteorology, which is essentially a science of observation and experiment, instruments are required for ascertaining, 1. the pressure of the atmosphere at any time or place ; 2. the temperature of the air ; 3. the absorption and radiation of the sun's heat by the earth's surface; 4. the humidity of the air ; 5. the amount and duration of rainfall ; 6. the direction, the horizontal pressure, and the velocity of winds; 7. the electric condition of the atmosphere, and the prevalence and activity of ozone.

CHAPTER I.

INSTRUMENTS FOR ASCERTAINING THE ATMOSPHERIC PRESSURE.

1. Principle of the Barometer.—The first instrument which gave the exact measure of the pressure of the atmosphere was invented by Torricelli, in 1643. It is constructed as follows :—A glass tube, C D (fig. 1), about 34 inches long, and from two to four-tenths of an inch in diameter of bore, having one end closed, is filled with mercury. In a cup, B, a quantity of mercury is also poured. Then, placing a finger securely over the open end, C, invert the tube vertically over the cup, and remove the finger when the end of the tube dips into the mercury. The mercury in the tube then partly falls out, but a column, A B, about 30 inches in height, remains supported. This column is a weight of mercury, the pressure of which upon the surface of that in the cup is precisely equivalent to the corresponding pressure of the atmosphere which would be exerted in its place if the tube were removed. As the atmospheric pressure varies, the length of this mercurial column also changes. It is by no means constant in its height; in fact, it is very seldom stationary, but is constantly rising or falling through a certain extent of the tube, at the level of the sea, near which the above experiment is supposed to be performed. It is, therefore, an instrument by which the fluctuations taking place in the

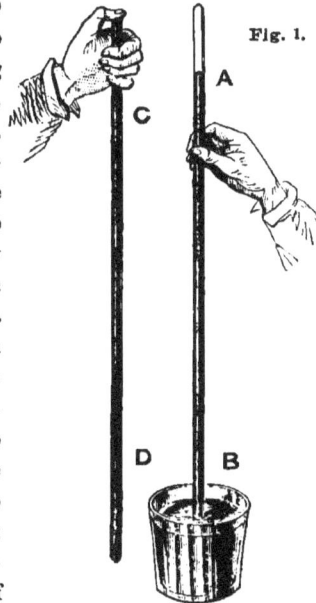

pressure of the atmosphere, arising from changes in its weight and elasticity, can be shown and measured. It has obtained the name *Barometer*, or measurer of heaviness,— a word certainly not happily expressive of the utility of the invention. If the bore of the barometer tube be uniform throughout its length, and have its sectional area equal to a square inch, it is evident that the length of the column, which is supported by the pressure of the air, expresses the number of cubic inches of mercury which compose it. The weight of this mercury, therefore, represents the statical pressure of the atmosphere upon a square inch of surface. In England the annual mean height of the barometric column, reduced to the sea-level and to the temperature of 32° Fahrenheit, is about 29·95 inches. A cubic inch of mercury at this temperature has been ascertained to weigh 0·48967 ℔s. avoirdupois. Hence, 29·95 × 0·48967= 14·67 ℔s., is the mean value of the pressure of the atmosphere on each square inch of surface, near the sea-level, about the latitude of 50 degrees. Nearer the equator this mean pressure is somewhat greater; nearer the poles, somewhat less. For common practical calculations it is assumed to be 15 ℔s. on the square inch. When it became apparent that the movements of the barometric column furnished indications of the probable coming changes in the weather, an attempt was made to deduce from recorded observations the barometric height corresponding to the most notable characteristics of weather. It was found that for fine dry weather the mercury in the barometer at the sea-level generally stood above 30 inches; changeable weather happened when it ranged from 30 to 29 inches, and when rainy or stormy weather occurred it was even lower. Hence, it became the practice to place upon barometer scales words indicatory of the weather likely to accompany, or follow, the movements of the mercury; whence the instruments bearing them obtained the name "Weather Glasses."

2. Construction of Barometers.—In order that the instrument may be portable, it must be made a fixture and mounted on a support; and, further, to render it scientifically or even practically useful, many precautions are required in its construction. The following remarks apply to the construction of all barometers:— Mercury is universally employed, because it is the heaviest of fluids, and therefore measures the atmospheric pressure by the shortest column. Water barometers have been constructed, and they require to be at least 34 feet long. Oil, or other fluids, might be used. Mercury, however, has other advantages: it has feeble volatility, and does not adhere to glass, if pure. Oxidised, or otherwise impure mercury, may adhere to glass; moreover, such mercury would not have the density of the pure metal, and therefore the barometric column would be either greater or less than it should be. The mercury of commerce generally contains lead; sometimes traces of iron and sulphur. It is necessary, therefore, for the manufacturer to purify the mercury; and this is done by washing it with diluted acetic, or sulphuric acid, which dissolves the impurities. No better test can be found for ascertaining if the mercury be pure than that of filling a delicate thermometer tube; if, on exhausting the air from this thermometer, the mercury will freely run up and down the bore, which is probably one thousandth of an inch in diameter, the mercury from which this thermometer was made will be found fit for any purpose, and with it a tube may be filled and boiled, not only of one inch, but even of two inches diameter. In all barometers it is requisite that the space above the mercurial column should be completely void of air and aqueous vapour, because these gases, by virtue of their elasticity, would depress the column. To exclude these the mercury is introduced, and boiled in the tube, over a charcoal fire, kept up for the purpose. In this manner the air and vapour which adhere to the glass are expanded, and escape away. One can tell whether a barometer has been properly "boiled," as it is termed, by simply holding the tube in a slanting direction and allowing the mercury to strike the top. If the boiling has been well performed, the mercury will give a clear, metallic sound ; if not, a dull, flat sound, showing some air to be present.

When the mercury in a barometer tube rises or falls, the level of the mercury in the cup, or *cistern*, as it is generally termed, falls or rises by a proportionate quantity, which depends upon the relative areas of the interior of the tube and of the cistern. It is necessary that this should be taken into consideration in ascertaining the exact height of the column. If a fixed scale is applied to the tube, the correct height may be obtained by applying a correction for capacity. A certain height of the mercury is ascertained to be accurately measured by the scale, and should be marked on the instrument as the *neutral point*. Above this point the heights measured are all less, and below, all more, than they should be. The ratio between the internal diameters of the tube and cistern (which should also be stated on the instrument, as, for instance, capac. 1/50) supplies the data for finding the correction to be applied. This correction is obviated by constructing the cistern so as to allow of the surface of the mercury in it being adjustable to the commence-ment of the fixed scale, as by Fortin's or Negretti's plan. It is also unnecessary in barometers constructed on what is now called the "Kew method." These will all be detailed in their proper place. The tube, being fixed to the cistern, may have a moveable scale applied to it. But such an arrangement requires the utmost care and skill in observing, and is seldom seen except in first-class Observatories.

3. Fortin's Barometer.—Fortin's plan of constructing a baro-meter cistern is shown in fig. 2. The cistern is formed of a glass cylinder, which allows of the level of the mercury within being seen. The bottom of the cylinder is made of sheep-skin or leather, like a bag, so as to allow of being pushed up or lowered by means of a screw, D B, worked from beneath. This screw moves through the bottom of a brass cylinder, C C, which is fixed outside, and protects the glass cylinder containing the mercury. At the top of the interior of the cistern is fixed a small piece of ivory, A, the point of which

Fig. 2.

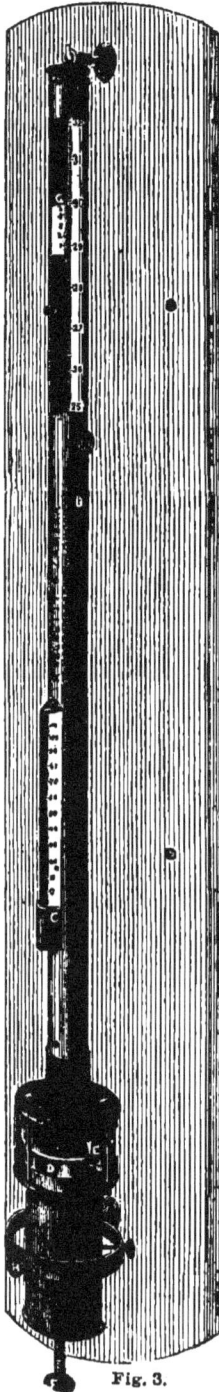

Fig. 3.

exactly coincides with the zero of the scale. This screw
and moveable cistern-bottom serve also to render the barometer portable, by confining the mercury in the tube, and
preventing its coming into the cistern, which is thus made
too small to receive it.

4. STANDARD BAROMETER.

Fig. 3 represents a Standard Barometer on Fortin's principle. The barometer tube is enclosed and protected by a
tube of brass extending throughout its whole length; the
upper portion of the brass tube has two longitudinal openings opposite each other; on one side of the front opening
is the barometrical scale of English inches, divided to show,
by means of a vernier, $\frac{1}{500}$th of an inch; on the opposite
side is sometimes divided a scale of French millimetres,
reading also by a vernier to $\frac{1}{10}$th of a millimetre (see
directions for reading the vernier, page 7). A thermometer, C, is attached to the frame, and divided to degrees,
which can be read to tenths; it is necessary for ascertaining
the temperature of the instrument, in order to correct the
observed height of the barometer.

As received by the observer, the barometer will consist
of two parts, packed separately for safety in carriage,—
1st, the barometer tube and cistern, filled with mercury,
the brass tube, with its divided scale and thermometer;
and 2nd, a mahogany board, with bracket at top, and brass
ring with three adjusting screws at bottom.

Directions for fixing the Barometer.—In selecting a position for a barometer, care should be taken to place it so
that the sun cannot shine upon it, and that it is not affected
by direct heat from a fire. The cistern should be from
two to three feet above the ground, which will give a height
for observing convenient to most persons. A standard
barometer should be compared with an observatory standard
of acknowledged accuracy, to determine its index error;
which, as such instruments are graduated by micrometrical
apparatus of great exactitude, will be constant for all parts
of the scale. It should be capable of turning on its axis
by a movement of the hand, so that little difficulty can ever
be experienced in obtaining a good light for observation.
Having determined upon the position in which to place the
instrument, fix the mahogany board as nearly vertical as
possible, and ascertain if the barometer is perfect and free

from air, in the following manner :—lower the screw at the bottom of the cistern several turns, so that the mercury in the tube, when held upright, may fall two or three inches from the top ; then slightly incline the instrument from the vertical position, and if the mercury in striking the top elicit a sharp tap, the instrument is perfect. Supposing the barometer to be in perfect condition, as it is almost sure to be, it is next suspended on the brass bracket, its cistern passing through the ring at bottom, and allowed to find its vertical position, after which it is firmly clamped by means of the three thumb-screws.

To Remove the Instrument when fixed to another Position.—If it should be necessary to remove the barometer,—first, by means of the adjusting screw, drive the mercury to the top of the tube, turning it gently when it is approaching the top, and cease directly any resistance is experienced ; next, remove from the upper bracket or socket ; lift the instrument and invert it, carrying it with its lower end upwards.

Directions for taking an Observation.—Before making an observation, the mercury in the cistern must be raised or lowered by means of the thumb-screw, F, until the ivory point, E, and its reflected image in the mercury, D, are just in contact ; the vernier is then moved by means of the milled head, until its lower termination just excludes the light from the top of the mercurial column ; the reading is then taken by means of the scale on the limb and the vernier. The vernier should be made to read upward in all barometers, unless for a special object, as this arrangement admits of the most exact setting. In observing, the eye should be placed in a right line with the fore and back edges of the lower termination of the vernier ; and this line should be made to form a tangent to the apex of the mercurial column. A small reflector placed behind the vernier and moving with it, so as to assist in throwing the light through the back slit of the brass frame on to the glass tube, is advantageous ; and the observer's vision may be further assisted by the aid of a reading lens. The object is, in these Standard Barometers, to obtain an exact reading, which can only be done by having the eye, the fore part of the zero edge of the vernier, the top of the mercurial column, and the back of the vernier, in the same horizontal plane.

Uniformity of Calibre.—The diameter of that part of the tube through which the oscillations of the mercury will take place is very carefully examined to insure uniformity of calibre, and only those tubes are used which are as nearly as possible of the same diameter throughout. The size of the bore should be marked on the frame of the barometer in tenths and hundredths of an inch. A correction due to capillary action, and depending on the size of the tube, must be applied to the readings.

5. Correction due to Capillarity.—When an open tube of small bore is plunged into mercury, the fluid will not rise to the same level inside as it has outside.

Hence, the effect of capillary action is to depress the mercurial column; and the more so the smaller the tube. The following table gives the correction for tubes in ordinary use :—

Diameter of tube.	Depression, in boiled tubes.	Depression, in unboiled tubes.
INCH.	INCH.	INCH.
0·60	0·002	0·004
0·55	0·003	0·005
0·50	0·008	0·007
0·45	0·005	0·010
0·40	0·007	0·015
0·35	0·010	0·021
0·30	0·014	0·029
0·25	0·020	0·041
0·20	0·029	0·058
0·15	0·044	0·086
0·10	0·070	0·140

This correction is always additive to the observed reading of the barometer.

6. Correction due to Temperature.—In all kinds of mercurial barometers attention must be given to the temperature of the mercury. As this metal expands and contracts very much for variations of temperature, its density alters correspondingly, and in consequence the height of the barometric column also varies. To ascertain the temperature of the mercury, a thermometer is placed near the tube, and is sometimes made to dip into the mercury in the cistern. The freezing point of water, 32°F., is the temperature to which all readings of barometers must be reduced, in order to make them fairly comparable. The reduction may be effected by calculation, but the practical method is by tables for the purpose; and for these tables we refer the reader to the works mentioned at the end of this book.

7. Correction due to Height above the Half-tide Level.—Further, in order that barometrical observations generally may be made under similar circumstances, the readings, corrected for capacity, capillarity, and temperature, should be reduced to what they would be at the sea-level, by adding a correction corresponding to the height above the mean level of the sea, or of half-tide. For practical purposes of comparison with barometric pressure at other localities, add one-tenth of an inch to the reading for each hundred feet of elevation above the sea. For scientific accuracy this will not suffice, but a correction must be obtained by means of Schuckburg's formula, or tables computed therefrom.

8. The Barometer Vernier.—The *vernier*, an invaluable contrivance for measuring small spaces, was invented by Peter Vernier, about the year 1630. The barometer scale is divided into inches and tenths. The vernier enables us to accurately subdivide the tenths into hundredths, and, in first-class instruments, even

Fig. 4. Fig. 5.

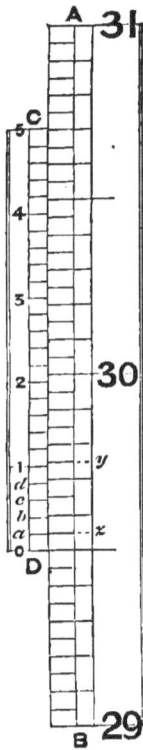

to thousandths of an inch. It consists of a short scale made to pass along the graduated fixed scale by a sliding motion, or preferably by a rack-and-pinion motion, the vernier being fixed on the rack, which is moved by turning the milled head of the pinion. The principle of the vernier, to whatever instrumental scale applied, is that the divisions of the moveable scale are to those in an equal length of the fixed scale in the proportion of two numbers which differ from each other by unity.

The scales of standard barometers are usually divided into half-tenths, or ·05, of an inch, as represented, in fig. 5, by A B. The vernier, C D, is made equal in length to twenty-four of these divisions, and divided into twenty-five equal parts; consequently one space on the scale is larger than one on the vernier by the twenty-fifth part of ·05, which is ·002 inch, so that such a vernier shows differences of ·002 inch. The vernier of the figure reading upwards, the lower edge, D, will denote the top of the barometer column; and is the zero of the vernier scale. In fig. 4, the zero being in line exactly with 29 inches and five-tenths of the fixed scale,

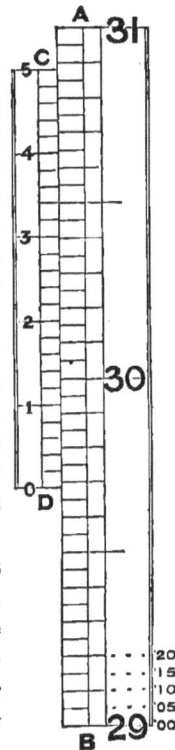

the barometer reading would be 29·500 inches. It will be seen that the vernier line, a, falls short of a division of the scale by, as we have explained, ·002 inch; b, by ·004; c, by ·006; d, by ·008; and the next line by one hundredth. If, then, the vernier be moved so as to make a coincide with z, on the scale, it will have moved through ·002 inch; and if 1 on the vernier be moved into line with y on the scale, the space measured will be ·010. Hence, the figures 1, 2, 3, 4, 5 on the vernier measure hundredths, and the intermediate lines even thousandths of an inch. In fig. 5, the zero of the vernier is intermediate 29·65 and 29·70 on the scale. Passing the eye up the vernier and scale, the second line above 3 is perceived to lie evenly with a line of the scale. This gives ·03 and ·004 to add to 29·65, so that the actual reading is 29·684 inches. It may happen that no line on the vernier accurately lies in the same straight line with one on the scale; in such a case a doubt will arise as to the selection of one from two equally coincident, and the intermediate thousandth of an inch should be taken.

For the ordinary purposes of the barometer as a "weather-glass," such minute measurement is not required. Hence, in household and marine barometers the scale need only be divided to tenths, and the vernier constructed to measure hun-

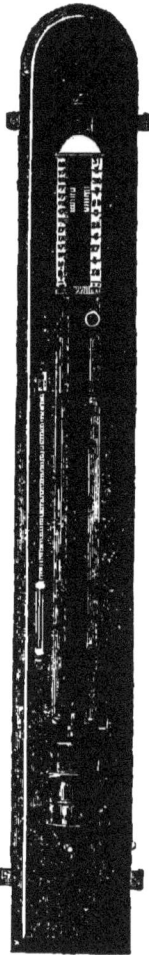

dredths of an inch. This is done by making the vernier either 9 or 11-10ths of an inch long, and dividing it into ten equal parts. The lines above the zero line are **Fig. 6.** then numbered from 1 to 10; sometimes the alternate divisions only are numbered, the intermediate numbers being very readily inferred. Hence, if the first line of the vernier agrees with one on the scale, the next must be out one-tenth of a tenth, or ·01 of an inch from agreement with the next *scale* line; the following vernier line must be ·02 out, and so on. Consequently, when the vernier is set to the mercurial column, the difference shown by the vernier from the tenth on the scale is the hundredths to be added to the inches and tenths of the scale.

A little practice will accustom a person to set and read any barometer quickly; an important matter where accuracy is required, as the heat of the body, or the hand, is very rapidly communicated to the instrument, and may vitiate, to some extent, the observation.

9. SELF-COMPENSATING STANDARD BAROMETER.

This barometer has been suggested to Messrs. Negretti and Zambra by Wentworth Erk, Esq. It consists of a regular barometer; but attached to the vernier is a double rack worked with one pinion, so that in setting or adjusting the vernier in one position, the second rack moves in directly the opposite direction, carrying along with it a plug or plunger the exact size of the internal diameter of the tube dipping in the cistern, so that whatever the displacement that has taken place in the cistern, owing to the rise or fall of the mercury, it is exactly compensated by the plug being more or less immersed in the mercury, so that no capacity correction is required.

A barometer on this principle is, however, no novelty, for at the Royal Society's room a very old instrument may be seen reading somewhat after the same manner.

Fig. 6 is an illustration of the appearance of this instrument. The cistern is so constructed that the greatest amount of light is admitted to the surface of the mercury.

10. BAROMETER WITH ELECTRICAL ADJUSTMENT.

This barometer is useful to persons whose eyesight may be defective; and is capable of being read off to greater accuracy than ordinary barometers, as will be seen by the following description:—The barometer consists of an upright tube dipping into a cistern, so contrived, that an up-and-down movement, by means of a screw, can be imparted to it. In the top of the tube a piece of platina wire is hermetically sealed. The cistern also has a metallic connection, so that by means of covered copper wires (in the back of the frame) a circuit is established; another connection also exists by means of a metallic point dipping into the cistern. The circuit, however, can be cut off from this by means of a switch placed about midway up the frame; on one side of the tube is placed a scale of inches; a small circular vernier, divided into 100 parts, is connected with the dipping point, and works at right angles with this scale.

To set the instrument in action for taking an observation, a small battery is connected by means of two small binding screws at the bottom of the frame. The switch is turned upwards, thereby disconnecting the dipping point; the cistern is then screwed up, so that the mercury in the tube is brought into contact with the platina wire at the top; the instant this is effected the magnetic needle seen on the barometer will be deflected. The switch is now turned down; by so doing the connection with the upper wire or platina is cut off, and established instead only between the dipping point carrying the circular vernier and the bottom of the cistern; the point is now screwed by means of the milled head until the needle is again deflected. We may now be sure that the line on the circular vernier that cuts the division on the scale is the exact height of the barometer. Although the description here given may seem somewhat lengthy, the operation itself is performed in less time than would be taken in reading off an ordinary instrument.

11. PEDIMENT BAROMETERS.

These Barometers, generally for household purposes, are illustrated by figs. 7 to 11.

Fig. 7.   Fig. 8.   Fig 9.   Fig. 10.   Fig. 11.

They are intended chiefly for "weather glasses," and are manufactured to serve not only a useful, but an ornamental purpose as well. They are usually framed in wood, such as mahogany, rosewood, ebony, oak or walnut, and can be obtained either plain or handsomely and elaborately carved and embellished, in a variety of designs, so as to be suitable for private rooms, large halls, or public buildings. The scales to the barometer and its attached thermometer may be ivory, porcelain, or silvered metal. It is not desirable that the vernier should read nearer than one-hundredth of an inch. Two verniers and scales may be fitted one on either side of the mercurial column, so that one can denote the last reading, and thus show at a glance the extent of rise or fall in the interval. The scale and thermometer should be covered with plate glass. A cheap instrument has an open face and plain frame, with sliding vernier instead of rack-and-pinion motion. The barometer may or may not have a moveable bottom to the cistern, with screw for the purpose of securing the mercury for portability. The cistern should not, however, require adjustment to a zero or fiducial point. It should be large enough to contain the mercury, which falls from 31 to 27 inches, without any appreciable error on the height read off on the scale.

12. The Words on the Scale.—The following words are usually engraved on the scales of these barometers, although they are not now considered of so much importance as formerly:—

At 31 inches	...	...	...	Very dry.
,, 30·5 ,,	...	...	...	Settled fair.
,, 30 ,,	...	...	...	Fair.
,, 29·5 ,,	...	...	...	Changeable.
,, 29 ,,	...	...	...	Rain.
,, 28·5 ,,	...	...	...	Much rain.
,, 28 ,,	...	...	...	Stormy.

The French place upon their barometers a similar formula :—

At 785 millimètres	...	...	Très-sec.
,, 776 ,,	...	...	Beau-fixe.
,, 767 ,,	...	...	Beau temps.
,, 758 ,,	...	...	Variable.
,, 749 ,,	...	...	Pluie ou vent.
,, 740 ,,	...	...	Grande pluie.
,, 731 ,,	...	...	Tempête.

Manufacturers of barometers have uniformly adopted these indications for all countries, without regard to the elevation above the sea, or the different geographical conditions; and as it can readily be shown that the height and variations of the barometer are dependent on these, it follows that barometers have furnished indications which, under many circumstances, have been completely false. Even in this country, and near the sea-level, storms are frequent with the barometer not below

29; rain is not uncommon with the glass at 30; even fine weather sometimes occurs with a low pressure; while it is evident that at an elevation of a few thousand feet the mercury would never rise to 30 inches; hence, according to the scale, there should never be fair weather there. If tempests happened as seldom in our latitude as the barometer gets down to 28 inches, the maritime portion of the community at least would be happy indeed. These words have long been ridiculed by persons acquainted with the causes of the barometric fluctuations; nevertheless opticians continue to place them on the scales, evidently only because they appear to add to the importance of the instrument in the eyes of those who have not learned their general inutility. In different regions of the world, the indications of the barometer are modified by the conditions peculiar to the geographical position and elevation above the sea, and it is necessary to take account of these in any attempt to found rules of general utility in connection with the barometer as a weather guide. All that can be said in favour of these words is, that within a few hundred feet of the sea-level, when the column rises or falls gradually during two or three days towards "Fair" or "Rain," the indications they afford of the coming weather are generally extremely probable; but when the variations are quick, upward or downward, they presage unsettled or stormy weather.

Admiral FitzRoy writes :—" The words on the scales of barometers should not be so much regarded, for weather indications, as the rising or falling of the mercury; for if it stands at *Changeable*, and then rises a little towards *Fair*, it presages a change of wind or weather, though not so great as if the mercury had risen higher; and, on the contrary, if the mercury stands above *Fair* and falls, it presages a change, though not to so great a degree as if it had stood lower; besides which, the direction and force of wind are not in any way noticed. It is not from the point at which the mercury stands that we are alone to form a judgment of the state of the weather, but from its *rising* or *falling;* and from the movements of immediately preceding days as well as hours, keeping in mind effects of change of *direction* and dryness, or moisture, as well as alteration of force or strength of wind."*

13. Correction due to Capacity of Cistern.—These barometers, having no adjustment for the zero of the scale, require a correction for the varying level of the mercury in the cistern, when the observations are required for strict comparison with other barometric observations, or when they are registered for scientific purposes; but for the common purpose of predicting the weather, this correction is unnecessary. The neutral point, and the ratio of the bore of the tube to the diameter of the cistern, must be known (see p. 3). Then the capacity correction, as it is termed, is found as follows:—Take the fractional part, expressed by the capacity ratio, of the difference between the observed reading and the height of the neutral point; then, if the mercury stand *below* the neutral point, *subtract* this result from the reading; if it stand *above*, *add* it to the reading.

---

* Second Number of *Meteorological Papers*, issued by the Board of Trade.

For example, suppose the neutral point to be 29·95 inches, and the capacity ratio $\frac{1}{50}$, required the correction when the barometer reads 30·78.

$$\text{Here} \qquad \qquad \qquad 30\cdot78 - 29\cdot95 = \quad 0\cdot83$$
$$\text{Correction} \qquad \qquad = \frac{0\cdot83}{50} = +\, 0\cdot02 \text{ nearly.}$$
$$\text{Scale reading} \qquad \qquad \qquad 30\cdot78$$
$$\text{Correct reading} \qquad \qquad \qquad \underline{30\cdot80}$$

Of course the correction could as easily be found to three decimal places, if desirable. It is evident that the correction is more important the greater the distance of the top of the mercury from the neutral point.

Fig. 12.

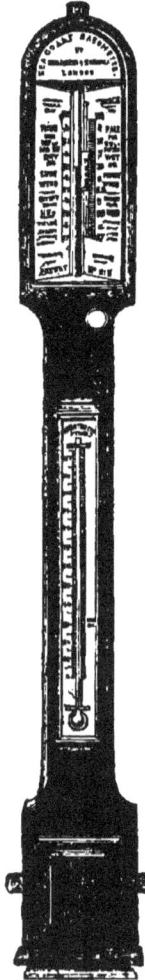

14. PUBLIC BAROMETERS.

Since the increased attention paid to the signs of forthcoming weather of late years, and the good which has resulted therefrom to farmers, gardeners, civil engineers, miners, fishermen, and mariners generally, by forewarning of impending wet or stormy weather, the desirability of having good barometers exposed in public localities has become evident.

Barometers may now be seen attached to drinking fountains, properly protected, and are frequently consulted by the passers-by. But it is among those whose lives are endangered by sudden changes in the weather, fishermen especially, that the warning monitor is most urgently required. Many poor fishing villages and towns have therefore been provided by the Board of Trade, at the public expense, and through the humane effort of Admiral FitzRoy, with first-class barometers, each fixed in a conspicuous position, so as to be easily accessible to all who desire to consult it. Following this example, the Royal National Life Boat Institution has supplied each of its stations with a similar storm warner ; the Duke of Northumberland and the British Meteorological Society have erected several on the coast of Northumberland ; and many other individuals have presented barometers to maritime places with which they are connected.

These barometers have all been manufactured by Messrs. Negretti & Zambra. The form given to the instrument seems well adapted for public purposes.

15. Fishery or Sea-coast Barometers.—Fig. 12 gives a representation of these coast and fishery barometers. The frame is of solid oak, firmly screwed together. The scales are very legibly engraved on porcelain by Negretti and Zambra's patent process. The thermometer is large, and easily read; and as this instrument is exposed, it will indicate the actual temperature sufficiently for practical purposes. The barometer tube is three-tenths of an inch in diameter of bore, exhibiting a good column of mercury; and the cistern is of such capacity, in relation to the tube, that the change of height in the surface of the mercury in the cistern corresponding to a change of height of three inches of mercury in the tube, is less than one-hundredth of an inch, and therefore, as the readings are only to be made to this degree of accuracy, this small error is of no importance. The cistern is made of boxwood, which is sufficiently porous to allow the atmosphere to influence the mercurial column; but the top is plugged with porous cane, to admit of free and certain play.

16. Admiral FitzRoy's Scale Words.—The directions given on the scales of these barometers were drawn up by Admiral FitzRoy, F.R.S. They appear to be founded on the following considerations:—

Supposing a compass diagram, with the principal points laid down, the N.E. is the wind for which the barometer stands highest; for the S.W. wind it is lowest. This is found to be so in the great majority of cases; but there are exceptions to this, as to all rules. The N.E. and S.W. may therefore be regarded as the poles of the winds, being opposite each other. When the wind veers from the S.W. through W. and N. to N.E., the barometer gradually rises; on the contrary, when the wind veers from N.E. and E. to S.E., S. and S.W., the mercury falls. A similar curious law exists in relation to the veering of the wind, and the action of the thermometer. As the wind veers from the S.W. to W. and N., the thermometer falls; as it veers from N.E. to E. and S., it rises, because the wind gets from a colder to a warmer quarter. The polar winds are cold, dry, and heavy. Those from the equatorial regions are warm, moist, and comparatively light.

These laws have been clearly developed and expressed by Professor Dové in his work on the "Law of Storms." The warm winds of Europe are those which bring the greatest quantity of rain, as they blow from the ocean, and come heavily laden with moisture. The cold winds, besides containing less moisture, blow more from the land. The weight of the vapour of the warm winds tends to raise the barometric column; but, at the same time, the increased dilatation of the air tends to lower it. This latter influence being the stronger, the barometer always falls for these winds; and in regions where they traverse a large extent of land, retain their heat, and become necessarily very dry, the fall in the barometer will be greater. Admiral FitzRoy's words for the scales of barometers for use in northern latitudes, then, are as follows:—

RISE.	*FALL.*
FOR	FOR
N. ELY.	S. WLY.
NW.--N.--E.	SE.--S.--W.
DRY	WET
OR	OR
LESS	MORE
WIND.	WIND.
EXCEPT	EXCEPT
WET FROM	WET FROM
N. ED.	N. ED.
Long foretold, long last;	First rise after low,
Short notice, soon past.	Foretells stronger blow.

It will be perceived that the exception in each case applies to N.E. winds. The barometer may fall with north-easterly winds, but they will be violent and accompanied with rain, hail, or snow; again, it will rise with these winds accompanied with rain, when they are light, and bring only little rain. It rises, however, highest with the dry and light N.E. winds.

These directions are very practically useful; they provide for geographical position—also for elevation above the sea—since they are not appended to any particular height of the column. They are suited to the northern hemisphere generally, as well as around the British Isles. The same directions are adapted to the southern hemisphere, by simply substituting for the letter N the letter S, reading south for north, and *vice versa*. South of the equator the cold winds come from the south; the warm, from the north. The S.E. wind in the southern hemisphere corresponds to the N.E. in the northern. The laws there are, while the wind veers from S.E. through E. to N. and N.W.; the barometer falls and the thermometer rises. As the wind veers from N.W. through W. and S. to S.E., the barometer rises and the thermometer falls.

17. Instructions for the Sea-coast Barometer.—The directions for fixing the barometer, and making it portable when it has to be removed, should be attended to carefully. The barometer should be suspended against a frame or piece of wood, so that light may be seen *through* the tube. Otherwise a piece of paper, or a *white place*, should be behind the upper or *scale part* of the *tube*.

When suspended on a hook, or stout nail, apply the milled-head key (which will be found just below the scales) to the square brass pin at the lower end of the instrument, and turn *gently* toward the left hand till the screw stops; then take off the key and replace it for use, near the scale, as it was before. The cistern bottom being thus *let down*, the mercury will sink to its proper level quickly.

In removing this barometer it is necessary to *slope it gradually*, till the mercury is at the top of the tube, and then, with the instrument reversed, to screw up the cistern bottom, or bag, by the key, used *gently*, till it stops. It will then be portable, and may be carried with the *cistern* end *upwards*, or lying flat; but it must not be jarred, or receive a concussion.

18. French Sea-coast Barometer.—The French have imitated this form of barometer for coast service, and have translated Admiral FitzRoy's indications for the scale as follows :—

<table>
<tr><td align="center">LA
HAUSSE
INDIQUE.</td><td align="center">LA
BAISSE
INDIQUE.</td></tr>
<tr><td align="center">DES VENTS DE LA
PARTIE DU
N.E.
(DU N.O. á l'E
PAR LE NORD.)</td><td align="center">DES VENTS DE LA
PARTIE DU
S.O.
(DU S.E. á l'O.
PAR LE SUD.)</td></tr>
<tr><td align="center">DE LA
SÉCHERESSE.</td><td align="center">DE
. L'HUMIDITÉ.</td></tr>
<tr><td align="center">UN VENT
PLUS FAIBLE
EXCEPTÉ S'IL PLEUT
AVEC DE FORTES BRISES
DU N.E.</td><td align="center">UN VENT
PLUS FORT
EXCEPTÉ S'IL PLEUT
AVEC DE PETITES BRISES
DU N.E.</td></tr>
<tr><td align="center">Mouvements lents,
Temps durable.</td><td align="center">Le commencement
de la hausse,
après une grande
baisse présage
un Vent violent.</td></tr>
<tr><td align="center">Mouvements rapides,
Temps variable.</td><td align="center"></td></tr>
</table>

Fig. 13.

Fig. 14.

MARINE BAROMETERS.

19. The Common Form.—The barometer is of great use to the mariner, who, by using it as a "weather glass," is enabled to foresee and prepare for sudden changes in the weather. For marine purposes, the lower portion of the glass tube of the barometer must be contracted to a fine bore, to prevent oscillation in the mercurial column, which would otherwise be occasioned by the movements of the ship. This tube is cemented to the cistern, which is made of boxwood, and has a moveable leathern bottom, for the purpose of rendering the instrument portable, by screwing up the mercury compactly in the tube. The tube is enclosed in a mahogany frame, which admits of a variety of style in shape, finish, and display, to meet the different fancies and means of purchasers. The frame is generally enlarged at the upper part to receive the scales and the attached thermometer, which are covered by plate glass. The cistern is encased in brass for protection, the bottom portion unscrewing to give access to the portable screw beneath the cistern. Figs. 13 and 14 illustrate this form of barometer. Marine barometers require to be suspended, so that they may remain in a vertical position under the changeable positions of a vessel at sea. To effect this they are suspended in gimbals by a brass arm. The gimbals consist of a loose ring fastened by thumb-screws to the middle part of the frame of the barometer, in front and back. The forked end of the arm supports this ring at the sides, also by the aid of thumb-screws. Hence the superior weight of the cistern end is always sufficient to cause the instrument to move on its bearing screws, so as always to maintain a per-pendicular position; in fact, it is so delicately held that it yields to the slightest disturbance in any direction. The other end of the arm is attached to a stout plate, having holes for screws, or fitted to slip into a staple or bracket, by which it may be fixed to any part of the cabin of a ship; the arm is hinged to the plate, for the pur-pose of turning the arm and barometer up whenever it is desirable.

Other forms of barometer (to be immediately described) have superseded this in the British Marine, but the French still give the preference to the wooden frames. They think the barometer can be more securely mounted in wood, is more portable, and less liable to be broken by a sudden concussion than if fitted in a metal frame. The English deem the ordinary wooden barometers not sufficiently accurate, owing to the irregular expansion of wood, arising from its hygrometric properties. Some of the English opticians have shown that very portable, and really accurate barometers can be made in brass frames, and therefore the preference is now given to this latter material.

20. The Kew Marine Barometer.—The form of barometer so-called, is that recommended by the Congress of Brussels, held in 1853, for the purpose of devising a systematic plan of promoting meteorological observations at sea.

The materials employed in its construction are mercury, glass, iron, and brass. The upper part of the tube is carefully calibrated to ensure uniformity of bore, as this is a point upon which the accuracy of the instrument to some extent depends. At sea, the barometer has never been known to stand above 31 inches, nor below 27. These extremes have been attained with instruments of undoubted accuracy, but they are quite exceptional. It is not necessary, therefore, to carry the scales of marine barometers beyond these limits, but they should not be made shorter. If the vernier is adjusted to read upward, the scale should extend to 32 inches, to allow room for the vernier to be set to 31 inches at least. Cases have occurred in which this could not be done, and rare, but valuable observations have been lost in conse-quence. If the scale part of the tube be not uniform in bore, the index error will be irregular throughout the scale. Whether the bore of the rest of the tube varies in diameter, is of no moment. From two to three inches below the measured part, the bore is contracted very much, to prevent the pulsations in the mercurial column—called "pumping"—which, otherwise, would occur at sea from the motion of the ship. In ordinary marine barometers, this contraction extends to the end of the tube. Below the contracted part is inserted a pipette—or Gay Lussac air-trap —which is a little elongated funnel with the point downwards. Its object is to arrest any air that may work in between the glass and the mercury. The bubble of air lodges at the shoulder, and can go up no farther. It is one of those simple contrivances which turn out remarkably useful. If any air gets into the tube, it does not get to the top, and therefore does not vitiate the performance of the barometer; for the mercury itself works up and down through the funnel. Below this, the tube should not be unnecessarily contracted.

The open end of the tube is fixed into an iron cylinder, which forms the cistern of the barometer. Iron has no action upon mercury, and is therefore used instead of any other metal. One or two holes are made in the top of the cistern, which are covered on the inside with strong sheep-skin leather, so as to be impervious to mercury, but sufficiently porous for the outer air to act upon the column. The cistern is of capacity sufficient to receive the mercury which falls out of the tube until the column stands lower than the scale reads; and when the tube is completely full, there is enough mercury to cover the extremity so as to prevent access of air. There is no screw required for screwing up the mercury.

Fig. 15.

The glass tube thus secured to the cistern is protected by a brass tubular frame, into which the iron cistern fits and screws compactly. Cork is used to form bearings for the tube. A few inches above the cistern is placed the attached thermometer. Its bulb is enclosed in the frame, so as to be equally affected by heat with the barometric column. The upper end of the frame is fitted with a cap which screws on, and embraces a glass shield which rests in a gallery formed on the frame below the scale, and serves to protect the silvered scale, as well as the inner tube, from dust and damp. A ring, moveable in a collar fixed on the frame above the centre of gravity of the instrument, is attached to gimbals, and the whole is supported by a brass arm in the usual manner; so that the instrument can be moved round its axis to bring any source of light upon it, and will remain vertical in all positions of the ship. The vernier reads to five-hundredths of an inch. No words are placed upon the scale, as the old formulary was deemed misleading. The vernier can be set with great exactness, as light is admitted to the top of the mercury by a front and a back slit in the frame. The lower edge of the vernier should be brought to the top of the mercury, so as just to shut out the light.

It is evident that this form of barometer must be more reliable in its indications than those in wooden frames. The graduations·can be accurately made, and they will be affected only by well-known alterations due to temperature. Some think the tube is too firmly held, and therefore liable to be broken by concussion more readily than that of an inferior instrument. This, however, appears a necessary conse-quence of greater exactness. It is an exceedingly good portable instrument, and can be put up and taken down very readily. These barometers are preferred to marine barometers in wood, wherever they have been used. In merchant ships, and under careful treatment, they have been found very durable. They may be sent with safety by railway, packed carefully in a wooden box.

Directions for Packing.—In removing this barometer it is necessary to slope it gradually till the mercury reaches the top of the tube. It is then portable, if carried cistern end upwards or lying flat. If carried otherwise, it will very probably be broken by the jerking motion of the heavy mercury in the glass tube. Of course it must not be jarred, or receive concussion.

Position for Marine Barometer.—Admiral FitzRoy, to whose valuable papers we are much indebted, writes in his "Barometer Manual" :—" It is desirable to place the barometer in such a position as not to be in danger of a side blow, and also sufficiently far from the deck above to allow for the spring of the metal arm in cases of sudden movements of the ship.

" If there is risk of the instrument striking anywhere when the vessel is much heeled, it will be desirable to put some soft padding on that place, or to check movement in that direction by a light elastic cord; in fixing which, attention must be paid to have it acting only where risk of a blow begins, not interfering otherwise with the free swing of the instrument: a very light cord attached above, when possible, will be least likely to interfere injuriously."

21. Method of verifying Marine and other Barometers.—"In nearly all the barometers which had been employed at sea till recently the index correction varied through the range of scale readings, in proportion to the difference of capacity between the cistern and the tube. To find the index correction for a land barometer, comparison with a standard, at any part of the scale at which the mercury may happen to be, is generally considered sufficient. To test the marine barometer is a work of much more time, since it is necessary to find the correction for scale readings at about each half inch throughout the range of atmospheric pressure to which it may be exposed ; and it becomes necessary to have recourse to artificial means of changing the pressure of the atmosphere on the surface of the mercury in the cistern.

" The barometers to be thus tested are placed, together with a standard, in an air-tight chamber, to which an air-pump is applied, so that, by partially exhausting the air, the standard can be made to read much lower than the lowest pressure to which marine barometers are likely to be exposed ; and by compressing the air it can be made to read higher than the mercury ever stands at the level of the sea. The tube of the standard is contracted similarly to that of the marine barometer, but a provision is made for adjusting the mercury in its cistern to the zero point. Glass windows are inserted in the upper part of the iron air-chamber, through which the scales of the barometers may be seen ; but as the verniers cannot be moved in the usual way from outside the chamber, a provision is made for reading the height of the mercury independent of the verniers attached to the scales of the respective barometers. At a distance of some five or six feet from the air-tight chamber a vertical scale is fixed. The divisions on this scale correspond exactly with those on the tube of the standard barometer. A vernier and telescope are made to slide on the scale by means of a rack and pinion. The telescope has two horizontal wires, one fixed and the other moveable by a micrometer screw, so that the difference between the height of the column of mercury and the nearest division on the scale of the standard, and also of all the other barometers placed by the side of it for comparison, can be measured either with the vertical scale and vernier or the micrometer wire. The means are thus possessed of testing barometers for index error in any part of the scale, through the whole range of atmospheric pressure to which they are likely to be exposed ; and the usual practice is to test them at every half inch from 27·5 to 31 inches.

" In this way barometers of various other descriptions have been tested, and some errors found to be so large that a few barometers read half an inch and upwards too high, while others read as much too low. In some cases those which were correct in one part of the scale were found to be from half an inch to an inch wrong in other parts. These barometers were of an old and ordinary, not to say inferior, construction. In some the mercury would not descend lower than about 29 inches, owing to a fault very general in the construction of many common barometers till lately in frequent use :—the *cistern was not large enough* to hold the mercury which descended from the tube in a *low atmospheric pressure.*

" When used on shore, this contraction of the tube causes the marine barometer to be *sometimes* a little behind an ordinary land barometer, the tube of which is not contracted. The amount varies according to the rate at which the mercury is rising or falling, and ranges from 0·00 to 0·02 of an inch. As the motion of the ship at sea causes the mercury to pass more rapidly through the contracted tube, the readings are almost the same there as they would be if the tube were not contracted, and in no case do they differ enough to be of importance in maritime use."

The cistern of this marine barometer is generally made an inch and a quarter in diameter, and the scale part of the tube a quarter of an inch in bore. The inches on the scale, instead of being true, are shortened by ·04 of an inch, in order to avoid the necessity of applying a correction due to the difference of capacity of the tube and cistern. This is done with much perfection, and the errors of the instruments, when compared with a standard by the apparatus used at Kew and

Liverpool Observatories, are determined to the thousandth of an inch, and are invariably very uniform and small.   The error so determined includes the correction due to capillarity, capacity, and error of graduation, and forms a constant correction, so that only one variable correction, that due to temperature, need be applied, when the barometer is suspended near the water line of the ship, to make the observations comparable with others.   With all the advantages of this barometer, however, it has recently been superseded, to some extent, because it was found to require more care than could ordinarily be expected to be given to it by the commander of a ship. Seamen do not exactly understand the value of such nice accuracy as the thousandth part of an inch, but prefer an instrument that reads only to a hundredth part.

22.   THE FITZROY MARINE BAROMETER.

Admiral FitzRoy deemed it desirable to construct a form of barometer as practically useful as possible for marine purposes.   One that should be less delicate in structure than the Kew barometer, and not so finely graduated.   One that could be set at a glance and read easily; that would be more likely to bear the common shocks unavoidable in a ship of war.   Accordingly, the Admiral has devised a barometer, which he has thus described:—

" This marine barometer, for Her Majesty's service, is adapted to *general* purposes.

" It differs from barometers hitherto made in points of detail, rather than principle:—1. The glass tube is packed with vulcanised india-rubber, which checks vibration from concussion; but does not hold it rigidly, or prevent expansion. 2. It does not oscillate (or pump), though extremely sensitive.   3. The scale is porcelain, *very legible*, and not liable to change.   4. There is no iron anywhere *(to rust)*.   5. Every part can be unscrewed, examined, or cleaned, by any careful person.   6. There is a *spare* tube, fixed in a cistern, filled with boiled mercury, and *marked* for adjustment in this, or *any similar* instrument.

" These barometers are graduated to hundredths, and they will be found accurate to *that* degree, namely the second decimal of an inch.

" They are packed with vulcanised caoutchouc, in order that (by this, and by a peculiar strength of glass tube) guns may be fired near these instruments without causing injury to them by ordinary concussion.

" It is hoped that all such instruments, for the public service at sea, will be quite similar, so that *any* spare tube will fit *any* barometer.

" *To Shift a Tube.*—Incline the barometer slowly, and then take it down, after allowing the mercury to fill the upper part.   Lay the instrument on a table, unscrew the outer cap at the joining just below the cistern swell, then unscrew the tube *and* cistern, by turning the cistern gently, against the sun, or to *the left*, and draw out the tube very carefully *without bending it in the least*, *turning* it a little, if required, as moved.   Then insert the new tube very cautiously, screw in, and adjust to the diamond-cut mark for 27 inches. Attach the cap, and suspend the barometer for use.

"If the mercury does not immediately quit the top of the tube, tap the cistern end rather sharply. In a well-boiled tube, with a good vacuum, the mercury hangs, at times, so adhesively as to deceive, by causing a supposition of some defect.

Fig. 16.

"In about ten minutes the mercurial column should be nearly right; but as local temperature affects the brass, as well as the mercury, slowly and unequally, it may be well to defer any *exact comparisons with other instruments* for some few hours."

Messrs. Negretti and Zambra are the makers of these barometers for the Royal Navy. Fig. 16 is an illustration.

The tube is fixed to a boxwood cistern, which is plugged with very porous cane at the top, to allow of the ready influence of a variation in atmospheric pressure upon the mercury. Round the neck of the cistern is formed a brass ring, with a screw thread on its circumference. This screws into the frame, and a mark on the tube is to be adjusted to 27 inches on the scale, the cistern covering screwed on, and the instrument is ready to suspend. The frame and all the fittings are brass, without any iron whatever; because the contact of the two metals produces a galvanic action, which is objectionable. The spare tube is fitted with india-rubber, and ready at any time to replace the one in the frame. The ease with which a tube can be replaced when broken is an excellent feature of the instrument. The spare tube is carefully stowed in a box, which can also receive the complete instrument when not in use. All the parts are made to a definite gauge; the frames are, therefore, all as nearly as possible similar to each other, and the tubes—like rifle bullets— are adjustible to any frame. If, then, the tube in use gets broken, the captain can replace it by the other; but, as it is securely packed with india-rubber, there is very little liability of its being broken by fair usage. Every person who knows the importance of the barometer on board ship, will acknowledge that the supplementary tube is a decided improvement. Many instruments of this description are afloat in the Royal Navy, and in a short time it may be expected that all the frames and tubes of barometers in the public service at sea will be similar in size and character; so that should a captain have the misfortune to get both his tubes broken, he would be able to borrow another from any ship he fell in with that had one to spare, which would be perfectly accurate, because it would have been verified before it was sent out.

23. Admiral FitzRoy's Words for the Scale.—The graduation of inches and decimals are placed in this barometer on the right-hand side of the tube; and on a

similar piece of porcelain, on the left-hand, are engraved, as legibly as they are
expressed succinctly, the following words, of universal application in the interpretation
of the barometer movements :—

RISE	*FALL*
FOR	FOR
COLD	WARM
DRY	WET
OR	OR
LESS	MORE
WIND.	WIND.
———	———
EXCEPT	EXCEPT
WET FROM	WET FROM
COOLER SIDE.	COOLER SIDE.

Reverting to the explanation of the words on the "Coast" barometers
(at page 14), and comparing and considering them as given for northern latitudes,
and as they must be altered for southern latitudes, it will be perceived, that for all
cold winds the barometer rises; and falls for *warm* winds. The mercury also falls
for *increased* strength of wind; and rises as the wind *lulls*. Likewise before or with
rain the column of mercury falls; but it rises with fine dry weather. Putting these
facts together, and substituting for the points of the compass the terms "cold" and
"warm," the appropriateness of the words on the scale of this barometer is readily
perceived. These concise and practical indications of the movements in the barometer are applicable for instruments intended for use in any region of the world,
and are in perfect accordance with the laws of winds and weather deduced by Dové
and other meteorologists. There is nothing objectionable in them, and being
founded upon experience and the deductions made from numerous recorded observations of the weather in all parts of the world, as well as confirmed by the theories
of science, they may consequently be considered as generally reliable. They involve
no conjecture, but express succinctly scientific principles.

24. Trials of the FitzRoy Marine Barometer under Fire of Guns.—
Some of the first barometers made by Messrs. Negretti and Zambra on Admiral
FitzRoy's principle were severely tried under the heaviest naval gun firing, on board
H.M.S. *Excellent;* and under all the circumstances, they withstood the concussion.
The purpose of the trials was "to ascertain whether the *vulcanized india-rubber
packing* round the glass tube of a *new marine barometer* did check the vibration
caused by firing, and whether guns might be fired close to these instruments
without causing injury to them." In the first and second series of experiments, a marine barometer on Admiral FitzRoy's plan was tried against a
marine barometer on the Kew principle, both instruments being new, and treated

in all respects similarly. They were "hung over the gun, under the gun, and by the side of the gun, the latter both inside and outside a bulkhead,—in fact, in all ways that they would be tried in action with the bulkheads cleared away." The result was that the Kew barometer was broken and rendered useless, while the new pattern barometer was not injured in the least. In a third series of experiments, Mr. Negretti being present, five of the new pattern barometers were subjected to the concussion produced by firing a 68-pounder gun with shot, and 16 ℔s. charge of powder. They were suspended from a beam immediately under the gun, then from a beam immediately over the gun, and finally they were suspended by the arm to a bulkhead, at a distance of only 3 ft. 6 in. from the axis of the gun; and the result was, according to the official report, "that all these barometers, however suspended, would stand, without the slightest injury, the most severe concussion that they would ever be likely to experience in any sea-going man-of-war." These trials were conducted under the superintendence of Captain Hewlett, C.B., and the guns were fired in the course of his *usual* instructions. His reports to Admiral FitzRoy, giving all the particulars of the trials, are published in the "Ninth Number of Meteorological Papers," issued by the Board of Trade.*

25. NEGRETTI AND ZAMBRA'S FARMER'S BAROMETER AND DOMESTIC WEATHER-GLASS.

It is a well-known fact that the barometer is as much, or even more affected by a change of wind as it is by rain; and the objection raised against a simple barometer reading, as leaving the observer in doubt whether to expect wind or rain, is removed by the addition of the Hygrometer, an instrument indicating the comparative degree of dryness or dampness of the air;—a most important item in the determination of the coming weather.

The farmer should not be content to let his crops lie at the mercy, so to speak, of the weather, when he has within his command instruments which may be the means of preventing damage to, and in cases total loss of, his crops.

The farmer hitherto has had to depend for his prognostication of the weather on his own unassisted "Weather Wisdom;" and it is perfectly marvellous how expert he has become in its use. Science now steps in, not to ignore this experience, but on the contrary, to give it most valuable assistance by extending it, and enabling it to predict, with an accuracy hitherto unknown, the various changes that take place in this most variable of climates.

To the invalid, the importance of predicting with tolerable accuracy the changes

---

* With reference to these barometers, we have received the subjoined testimonial, with permission to use it as we please.

"*Meteorologic Office*, 12th *June*, 1863.

"MESSRS. NEGRETTI & ZAMBRA,

"The barometers which you have lately supplied to Her Majesty's ships through this Office are much approved, being good for general service, afloat or on land.

"(Signed)   R. FITZROY."

that are likely to occur in the weather, cannot be over-rated. Many colds would be prevented, if we could know that the morning so balmy and bright, would subside into a cold and cheerless afternoon. Even to the robust, much inconvenience may be prevented by a due respect to the indications of the hygrometer and the barometer, and the delicate in health will do well to regard its warnings.

Fig. 17.

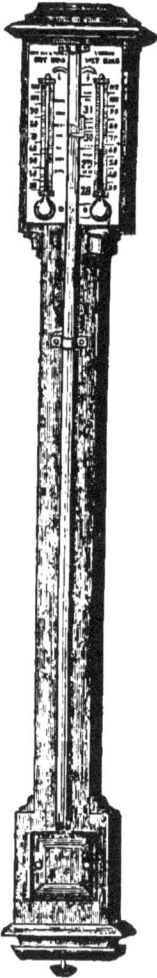

Description of the Instrument.—The farmer's barometer, as figured in the margin, consists of an upright tube of mercury inverted in a cistern of the same fluid; this is secured against a strong frame of wood, at the upper end of which is fixed the scale, divided into inches and tenths of an inch. On either side of the barometer, or centre tube, are two thermometers— that on the left hand has its bulb uncovered and freely exposed, and indicates the temperature of the air at the place of observation; that on the right hand has its bulb covered with a piece of muslin, from which depend a few threads of soft lamp cotton; this cotton is immersed in the small cup situated just under the thermometer, this vessel being full of water; the water rises by capillary attraction to the muslin-covered bulb, and keeps it in a constantly moist state.

These two thermometers, which we distinguish by the names "Wet Bulb" and "Dry Bulb," form the Hygrometer; and it is by the simultaneous reading of these two thermometers, and noting the difference that exists between their indications, that the humidity in the atmosphere is determined.

Admiral FitzRoy's words (see p. 22) are placed upon the scale of the barometer, as the value of a reading depends, not so much on the actual height of the mercury in the tube, as it does on whether the column is rising, steady, or falling.

The moveable screw at the bottom of the cistern is for the purpose of forcing the mercury to the top of the tube when the instrument is being carried from place to place, and it must always be unscrewed to its utmost limit when the barometer is hung in its proper place. After this it should never be touched.

The manner in which the Hygrometer acts is as follows: It is a pretty well-known fact that water or wine is often cooled by a wet cloth being tied round the bottle, and then being placed in a current of air. The evaporation that takes place in the progressive drying of the cloth causes the temperature to fall considerably below that of the surrounding atmosphere, and the contents of the bottle are thus cooled. In the same manner, then, the covered wet bulb thermometer will be found *invariably* to read lower than the uncovered one; and the greater the dryness of the air, the greater will be the difference between the indications of the two thermometers; and the more moisture that exists in the air, the more nearly they will read alike.

The cup must be kept filled with pure water, and occasionally cleaned out, to remove any dirt. The muslin, or cotton-wick, should also be renewed every few weeks. The hygrometer may be had separate from the barometer, if the combined instruments cannot be sufficiently exposed to the external air, this being essential for the successful use of the hygrometer.

This farmer's weather-glass, then, consists of three distinct instruments: the barometer, the thermometer, and the hygrometer. He has thus at command the three instrumental data necessary for the prediction of the weather. And now to describe—

How to Use the Instrument.—The observations should be taken twice a day, say at 9 A.M. and 3 P.M.; and should be entered on a slip of paper, or a slate hung up by the barometer. The observer will then be able to see the different values of the readings from time to time, and to draw his conclusions therefrom.

The thermometer on the left hand should first be read, and a note made of its indication, which is the temperature of the air. The wet bulb thermometer should now be read, and also noted; and the difference should be taken of these two readings. Next read the barometer by moving the small index at the side of the tube until it is on a level with the top of the mercury. Having noted the number of inches at which the column stands, compare with the last observation, and see immediately whether the barometer is rising, steady, or falling.

Now, having taken the observations as above, we naturally ask the question, *What are we to predict from them?*

And, probably, the best way of answering this query will be by giving an example. We will suppose that our readings yesterday were as follows :—Temperature, 70°; Wet Bulb, 69°; Difference, 1°;=very moist air. Barometer, 29·5, and that rain has fallen.

To-day, we read:—Temperature, 60°; Wet Bulb, 55°; Difference, 5°;=dryer air. Barometer, 30. We may safely predict that the rain will cease, and probably we may have wind from the northward.

In spring or autumn, if the barometric height be steady any where between 29·5 and 30 inches, with the temperature about 60°, fresh to moderate south-westerly winds, with cloudy sky, will probably characterize the weather; the indications of the hygrometer being then specially serviceable in enabling us to foretell rain; but if the mercury become steady at about 30·5 inches, with temperature about 40°, north-easterly winds, dry air, and clear sky, may be confidently expected.

Many cases will doubtless suggest themselves to the observer where these figures do not occur, and where he might find a difficulty in interpreting the indications of his instruments. We have, therefore, drawn up some concise rules for his guidance; and although they will not prove absolutely infallible guides to this acknowledged most difficult problem, still, they will be found of much service in foretelling the weather, when added to an intelligent observation of ordinary atmospheric phenomena, as force and direction of wind, nature of any particular season, and the time of year.

26. RULES FOR FORETELLING THE WEATHER.

A RISING BAROMETER.

A " Rapid " rise indicates unsettled weather.

A " Gradual " rise indicates settled weather.

A " Rise," with dry air, and cold increasing in summer, indicates wind from northward; and if rain has fallen, better weather is to be expected.

A " Rise," with moist air and a low temperature, indicates wind and rain from northward.

A " Rise," with southerly wind, indicates fine weather.

A STEADY BAROMETER,

With dry air and a seasonable temperature, indicates a continuance of very fine weather.

A FALLING BAROMETER.

A " Rapid " fall indicates stormy weather.

A " Rapid " fall, with westerly wind, indicates stormy weather from northward.

A " Fall," with a northerly wind, indicates storm, with rain and hail in summer, and snow in winter.

A " Fall," with increased moisture in the air, and the heat increasing, indicates wind and rain from southward.

A " Fall," with dry air, and cold increasing (in winter), indicates snow.

A " Fall," after very calm and warm weather, indicates rain with squally weather.

27, Causes which may bring about a Fall or a Rise in the Barometer.[*]— As heat produces rarefaction, a sudden rise of temperature in a distant quarter may affect the weight of the atmosphere over our heads, by producing an aerial current outwards, to supply the place of the lighter air which has moved from its former position; in which case the barometer will fall. Now such a movement in the atmosphere is likely to bring about an intermixture of currents of air of different temperatures, and from this intermixture rain is likely to result.

On the other hand, as cold produces condensation, any sudden fall of temperature causes the column of air over the locality to contract and sink to a lower level, whilst other air rushes in from above to supply the void; and, accordingly, the barometer rises. Should this air, as often happens, proceed from the north, it will contain in general but little moisture; and hence, on reaching a warmer latitude, wil take up the vapour of the air, so that dry weather will result.

It is generally observed, that wind causes a fall in the instrument; and, indeed, in those greater movements of the atmosphere which we denominate storms or

---

[*] *Vide* C. Daubeny, F.R.S., "On Climate."

hurricanes, the depression is so considerable as to forewarn the navigator of his impending danger. It is evident, that a draught of air in any direction must diminish the weight of the column overhead, and consequently cause the mercury in the barometer to sink.

The connection, therefore, of a sinking of the barometric column with rain is frequently owing to the wind causing an intermixture of the aerial currents which, by their motion, diminish the weight of the atmosphere over our heads; whilst a steady rise in the column indicates the absence of any great atmospheric changes in the neighbourhood, and a general exemption from those causes which are apt to bring about a precipitation of vapour.

28. Use of the Barometer in the management of Mines.—The inflammable and suffocating gases, known to coal-miners as fire-damp and choke-damp, are specifically heavier than air; and as they issue from the fissures of the mine, or are released from the coal, the atmospheric pressure tends to drive them into the lowest and least ventilated galleries. Consequently a greatly reduced atmospheric pressure will favour a sudden outflow or advance of gas; whence may result cases of explosion or suffocation. It has been found that these accidents occur for the most part about the time of a low barometric column. A reliable barometer should, therefore, be systematically consulted by those entrusted with the management or control of coal-mines, so that greater vigilance and caution may be enjoined on the miners whenever the mercury falls low, especially after it has been unusually high for some days.

29. Use of the Barometer in estimating the Height of Tides.—The pressure of the atmosphere affects the height of the tide, the water being in general higher as the barometer is lower. The expressions of seamen, that "frost nips the tide," and "fog nips the tide," are explained by the high barometer which usually accompanies frost and fog. M. Daussy, Sir J. C. Ross, and others, have established that a rise of one inch in the barometer will have a corresponding fall in the tide of about one foot. Therefore navigators and pilots will appreciate the following suggestion of Admiral FitzRoy:—

"Vessels sometimes enter docks, or even harbours, where they have scarcely a foot of water more than their draught; and as docking, as well as launching large ships, requires a close calculation of height of water, the state of the barometer becomes of additional importance on such occasions."

CHAPTER II.

SYPHON TUBE BAROMETERS.

30. Principle of.—If some mercury, or any other fluid, be poured into a tube of glass, bent in the form of U, and open at both ends, it will rise to the same height in both limbs, the tube being held vertically. If mercury be poured in first, and then water upon it at one end, these liquids will not come to the same level; the water will stand much higher than the mercury. If the height of the mercury, above the line of meeting of the fluids, be one inch, that of the water will be about thirteen-and-a-half inches. The explanation of this is, that the two columns balance each other. The pressure of the atmosphere in each limb is precisely similar; but the one column stands so much higher than the other, because the fluid of which it is composed is so much lighter, bulk for bulk, than the other. If one end of the tube be hermetically closed, the other limb be cut off within a few inches of the bend, and the tube carefully filled with mercury; by placing it in a vertical position, the mercury will fall, if the closed limb be long enough, until it is about thirty inches higher than that in the exposed limb, where it will remain. Here the atmosphere presses upon the short column; but not upon the long one. It is this pressure, therefore, which maintains the difference of level. In fact, it forms a barometer without a cistern, the short limb answering the purpose of a cistern. The first barometers on this principle were devised by the celebrated philosopher, Dr. Hook, as described in the next section.

31. DIAL, OR WHEEL BAROMETERS.

The familiar household "Weather Glasses" are barometers on the syphon principle. The portions of the two limbs through which the mercury will rise and fall with the varying pressure of the atmosphere are made of precisely the same diameter; while the part between them is contracted. On the mercury, in the exposed limb, rests a round float of ivory or glass; to this a string is attached and passed over and around a brass pulley, the other end carrying another lighter weight. The weight resting on the mercury rises and falls with it. On the spindle of the pulley, which passes through the frame and centre of the dial-plate, is fixed a light steel hand, which revolves as the pulley turns round. When the mercury falls for a decrease of atmospheric pressure, it rises by the same quantity in the short tube, and pushes up the float, the counterpoise falls, and thus moves the hand or pointer to the left. When the pressure increases, the pointer is drawn in a similar manner to the right.

Fig. 19.

Fig. 20.

Fig. 18.

The dials are generally made of metal silvered over or enamelled, but porcelain may be used. If the circumference of the pulley, or "wheel," be two inches, it will revolve once for an alteration of level amounting to two inches in each tube, or four inches in the height of the barometric column ; and as the dial may be from twenty to thirty-six inches in circumference, five to nine inches on the graduated scale corresponds to one inch of the column ; and hence the sub-divisions are distinctly perceptible, and a vernier is not necessary.

The motion of the pointer alone is visible ; and a mahogany, or rosewood, frame, supports, covers, and renders the instrument ornamental and portable. In the back of the frame is a hinged door, which covers the cavity containing the tube and fixtures. The dial is covered by a glass in a brass rim, similar to a clock face. A brass index, working over the dial, moveable by a key or button, may be applied, and will serve to register the position of the hand when last observed. These instruments are usually fitted with a thermometer, and a spirit level ; the latter for the purpose of getting the instrument perfectly vertical. They sometimes have, in addition, a hygrometer, a sympiesometer, an aneroid, a mirror, or a clock, &c., singly or combined. The frame admits of much variety of style and decoration. It may be carved or inlaid. The usual adjustment of scale is suited for localities at no considerable elevation above the sea. Accordingly, being commercial articles, they have been found frequently quite out of place. When intended for use at high

Fig. 22.

Fig. 21.

Fig. 23.

elevations, they should have a special adjustment of scale. As household instruments they are serviceable, and ornamental. But the supply-and-demand principle upon which they are sold, has entailed upon those issued by inferior makers a generally bad adjustment of scale. The illustrations are those of ordinary designs.

Dial barometers required for transmission to distant parts, as India and the Colonies, are furnished with a steel stop-cock, to render them portable more effectually than can be done by the method of *plugging* the tube.

32. STANDARD SYPHON BAROMETER.

Fig. 24 represents the most accurate form of the Gay Lussac barometer. The short limb is closed at the top, after the mercury is introduced, and a small lateral

Fig. 24.

puncture is made at *a*, which is covered over with a substance which permits the access of air, but prevents the escape of any mercury when the instrument is packed for travelling. The bent part of the tube is contracted to a capillary bore; and just above this, in the long limb, is placed the air-trap, already described (see p. 17), and here illustrated (fig. 25). When reversed, as it must be for portability, the capillary attraction keeps the mercury in the long branch. Should the mercury of the short column get detached, some small quantity of air *may* pass; but it will be arrested at the pipette, and will **Fig. 25.** not vitiate the length of the barometric column. It can be easily expelled by gently shaking or tapping the instrument before suspending it for observation. In the illustration, the zero of the scale is placed at Z, near the middle of the tube; and the graduations extend above and below. In making an observation, it is necessary to take the reading Z A on the long branch, and Z B on the short one. The sum of the two gives the height of the barometer. The zero of the scale in some instruments is placed low down, so as to require the difference of the two readings to be taken. A thermometer is attached to the frame as usual.

These instruments can be very accurately graduated, and are very exact in their indications, provided great care has been exercised in selecting the tubes, which must be of the same calibre throughout the parts destined to measure the variations of atmospheric pressure. They should be suspended so as to insure their hanging vertically.

The syphon barometer does not require correction for capillarity nor for capacity, as each surface of the mercury is equally depressed by capillary attraction, and the quantity of mercury which falls from the long limb of the tube occupies the same length in the short one. The barometric height must, however, be corrected for temperature, as in the cistern barometer. Tables containing the temperature corrections to be applied to barometer readings for scales engraved on the glass tube, or on brass or wood frames, are published.

CHAPTER III.

BAROGRAPHS, OR SELF-REGISTERING BAROMETERS.

33. Milne's Self-Registering Barometer.—For a long time a good and accurate self-recording barometer was much desired. This want is now satisfactorily supplied, not by one, but by several descriptions of apparatus. The one first to be described was the design of Admiral Sir A. Milne, who himself constructed, in 1857, we believe, the original instrument, which he used with much success. Since that time several of these instruments have been made, and have performed satisfactorily. The barometer tube is a syphon of large calibre, provided with a Gay Lussac pipette, or air-trap; and fitted with a float, a wheel, and a pointer, as in the "Dial" barometer. The float is attached to a delicate watch-chain, which passes over the wheel and is adequately counterpoised. Behind the indicating extremity of the pointer or hand is a projecting point, which faces the frame of the instrument, and is just within contact with the registering paper. A clock is applied, and fitted with auxiliary mechanism, so as to be able to move the mounted paper with regularity behind the pointer, and at designed equal intervals of time to release a system of levers and springs, so as to cause the marker to impress a dot on the paper, either by puncture or pencil-mark. The paper is ruled with horizontal lines for the range of the mercurial column, and parallel arcs of circles for the hours. Thus the barometer is rendered self-recording, by night or day, for a week or more; hence the great value of the instrument. The clock, index, and registering mechanism are protected from dust and interference by a glass front, hinged on and locked. As the temperature of the mercury is not registered, there is fixed to the frame a Sixe's thermometer to record the maximum and minimum temperatures, which should be noted at least every twenty-four hours.

Admiral FitzRoy has suggested the name "Atmoscope" for Admiral Milne's barometer; and he has also termed it a "Barograph." This latter word appears to be applicable to all kinds of self-registering barometers hitherto designed. Of the arrangement under consideration Admiral FitzRoy writes :—"It shows the alterations in tension, or the pulsations, so to speak, of atmosphere, on a large scale, by hourly marks ; and the diagram expresses, to a practised observer, what the 'indicator card' of a steam-cylinder shows to a skilful engineer, or a stethescope to a physician."

34. Modification of Milne's Barometer.—The great difficulty to be overcome in Milne's barometer, is to adjust the mechanism for obtaining registration so that the action of the striker upon the pointer should not in the slightest degree move it from its true position. A different mode of registration, capable of recording accurately the least appreciable movement of the mercurial column, has been effected. In this instrument the registering paper is carried upon a cylinder or drum. By reference to the illustration, Fig. 26, the details of construction will be readily understood. It should, however, be mentioned, that it is not a picture of the outward appearance of the instrument. The position of the barometer should be behind the clock; it is represented on one side merely for the purpose of clearly illustrating the arrangement and principles. The instrument has a large syphon barometer tube, in which the mercurial column is represented. On the mercury at A, in its open end, rests a glass float, attached to a watch-chain, or suitable silken cord, the other end of which is connected to the top of the arched head on the short arm of a lever-beam. The long arm of the beam is twice the length of the short arm, for the following reason. As the mercury falls in the long limb, it rises through an equal space in the short limb of the tube, and *vice versa*. But the barometric column is the difference of height of the mercury in the two limbs; hence the rise or fall of the float through half-an-inch will correspond to a decrease or an increase of the barometric column of one inch. In order, then, to record the movements of the barometric column, and not those of the float, the arm of the beam connected with the float is only half the radius of the other arm. Both arms of the beam carry circular-arched heads, which are similar portions of the complete circles, the centre of curvature being the fulcrum, or axis. This contrivance maintains the leverage on each extremity of the beam always at the same distance from the fulcrum. From the top of the large arched head a piece of watch-chain descends, and is attached to the marker, B, which properly counterpoises the float, A, and is capable of easy movement along a groove in a brass bar, so as to indicate the barometric height on an ivory scale, C, fixed on the same vertical framing. On the opposite side of the marker, B, is formed a metallic point, which faces the registration sheet and is nearly in contact with it. The framing, which carries the scale and marker, is an arrangement of brass bars, delicately adjusted and controlled by springs, so as to permit of a quick horizontal motion, in a small arc, being communicated to it by the action of the hammer, E, of the clock, whereby the point of the marker is caused to impress a

Fig. 26.

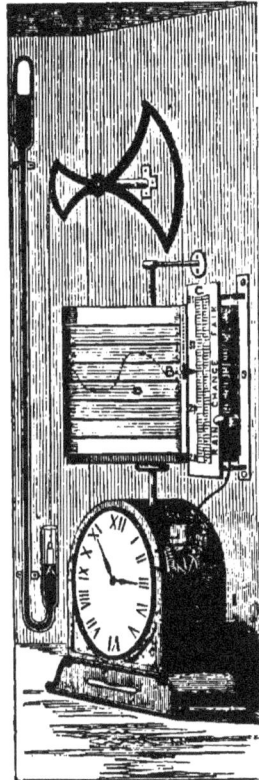

dot upon the paper. The same clock gives rotation to the hollow wooden cylinder, *D*, upon which is mounted the registering paper. The clock must be rewound when a fresh paper is attached to the cylinder, which may be daily, weekly, or monthly, according to construction; and the series of dots impressed upon the paper shows the height of the barometric column every hour by day and night. The space traversed by the marker is precisely equal to the range of the barometric column.

Fig. 27.

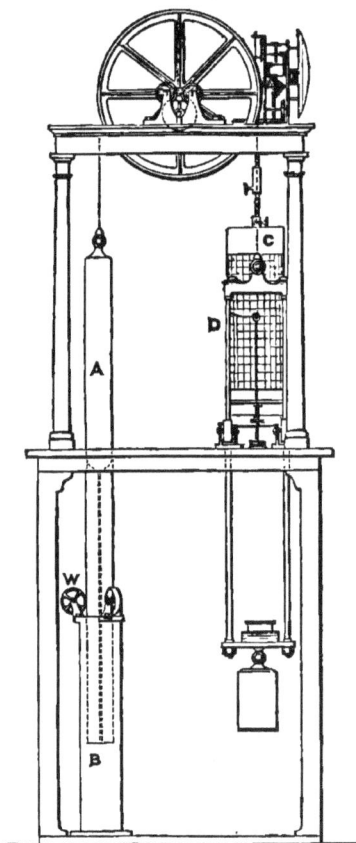

35. King's Self-Registering Barometer.

Mr. Alfred King, Engineer of the Liverpool Gas-light Company, designed, so long ago as 1854, a barometer to register, by a continuous pencil-tracing, the variations in the weight of the atmosphere; and a highly-satisfactory self-recording barometer, on his principle and constructed under his immediate superintendence, has quite recently been erected at the Liverpool Observatory.

Fig. 27 is the front elevation of this instrument. *A*, the barometer tube, is three inches in internal diameter, and it floats freely (not being fixed as usual) in the fixed cistern, *B*, guided by friction-wheels, *W*. The top end of the tube is fastened to a peculiar chain, which passes over a grooved wheel turning on finely-adjusted friction rollers. The other end of the chain supports the frame, *D*, which carries the tracing pencil. The frame is suitably weighted and guided, and faces the cylinder, *C*, around which the tracing paper is wrapped, and which rotates once in twenty-four hours by the movement of a clock. Mr. Hartnup, Director of the Liverpool Observatory, in his Annual Report, 1868, says:—"For one inch change in the mercurial column the pencil is moved through five inches, so that the horizontal lines on the tracing, which are half an inch apart, represent one-tenth of an inch change in the barometer. The vertical lines are hour lines, and being nearly three-quarters of an inch apart, it will be seen that the smallest appreciable change in the barometer, and the time of its occurrence, are recorded."

"It has been remarked by persons in the habit of reading barometers with large tubes, that, in squally weather, sudden and frequent oscillations of the mercurial column are sometimes seen.   Now, to register these small oscillations must be a very delicate test of the sensitiveness of a self-registering barometer, as the time occupied by the rise and fall of the mercury in the tube in some cases does not exceed one minute."   Mr. Hartnup affirms that the tracing of this instrument exhibits such oscillations whenever the wind blows strong and in squalls.

As the barometer in this instrument is precisely similar to the "Long Range Barometer" invented by Mr. McNeild (and which will be found described at page 48), it may be desirable to quote the following, from Mr. Hartnup's Report :—" Mr. King constructed a small model instrument to illustrate the principle. This instrument was entrusted to my care for examination, and it was exhibited to the scientific gentlemen who visited the Observatory in 1854, during the meeting of the British Association for the Advancement of Science."

36. Syphon, with Photographic Registration.—A continuous self-registering barometer has been constructed, in which photography is employed. Those who may wish to adopt a similar apparatus, or thoroughly to understand the arrangements and mode of observation, should consult the detailed description given in the *Greenwich Magnetical and Meteorological Observations*, 1847.   As the principles are applicable to photographic registration of magnetic and electric as well as meteorologic variations in instrumental indications, it would be beside our purpose to describe fully the apparatus.

The barometer is a large syphon tube ; the bore of the upper and lower extremities, through which the surfaces of the mercury rise and fall, is $1\frac{1}{10}$ inch in diameter.   The glass float in the open limb is attached to a wire, which moves a delicately-supported light lever as it alters its elevation.   The fulcrum of the lever is on one side of the wire; the extremity on the other side, at four times this distance from the fulcrum, carries a vertical plate of opaque mica, having a small aperture.   Through this hole the light of a gas-jet shines upon photographic paper wrapped round a cylinder placed vertically, and moved round its axis by a clock fixed with its face horizontal.   The cylinder is delicately supported, and revolves in friction rollers.   A bent wire on the axis is embraced by a prong on the hour hand of the time-piece; therefore the cylinder is carried round once in twelve hours.   It might be arranged for a different period of rotation.

As the cylinder rotates, the paper receives the action of the light, and a photographic trace is left of the movements of the barometer four times the extent of the oscillations of the float, or twice the length of the variations in the barometric column.   Certain chemical processes are required in the preparation of the paper, and in developing the trace.   The diagram which we give on the next page, with the explanation, taken from Drew's *Practical Meteorology*, will enable the above description to be better understood :

"$Q e$ is a lever whose fulcrum is e, the counterpoise f nearly supporting it; s is an opaque plate of mica, with a small aperture at p, through which the light passes, having before been refracted by a cylindrical lens into a long ray, the portion only of which opposite the aperture p impinges on the paper; d is a wire supported by a float on the surface of the mercury; $G H$ is the barometer; p, the vertical cylinder charged with photographic paper; r, the, photographic trace; I, the timepiece, carrying round the cylinder by the projecting arm t. It is evident that the respective distances of the float and the aperture p from the fulcrum may be regulated so that

Fig. 28.

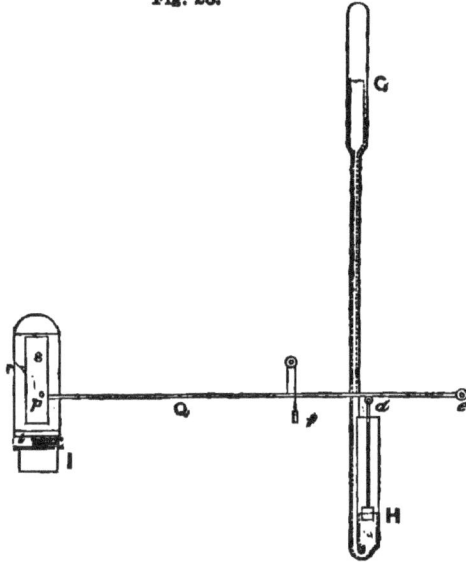

the rise and fall of the float may be multiplied to any extent required." When *only* the lower surface of the mercury in a syphon barometer is read, as in the instrument just described, a correction for temperature is strictly due to the height of the quicksilver in the *short* tube; but this in so short a column will rarely be sensible.

CHAPTER IV.

MOUNTAIN BAROMETERS.

37. The Syphon Tube Mountain Barometer, on Gay Lussac's principle, constructed as described at page 31, and fixed in a metallic tubular frame, forms a simple and light travelling instrument. The graduations are made upon the frame, and it is suspended for reading by a ring at the top, from beneath an iron tripod stand, which is usually supplied with it. Considerable care is requisite in adjusting the verniers, so as to keep the instrument steady and vertical. A drawback to the convenience of this barometer is the movement of the mercury in the short limb, which is generally not confined, and hence has every facility for becoming quickly oxidised in travelling. To remedy this, Messrs. Negretti and Zambra so construct the Mountain Syphon Barometer that by a simple half turn of a screw the mercury can be confined for portability, while the lower limb can be taken out for cleaning whenever found requisite.

38. Mountain Barometer on Fortin's principle.—This barometer, with Fortin's cistern, as arranged by Messrs. Negretti and Zambra, is an elegant, manageable, and very accurate instrument for travelling purposes, and well adapted for careful measurement of heights. The cistern is made large enough to receive all the mercury that will fall from the tube at the highest attainable elevation. The screw at the bottom confines the mercury securely for carriage, and serves to adjust the surface of the mercury to the zero of the scale when making an observation. The vernier reads to ·002 of an inch, and slides easily on the brass frame, which is made as small in diameter as is compatible with the size of the tube. The tube in this barometer should be altogether without contractions, so that the mercury will readily fall when it is set up for observation. It must be carefully calibrated, and its internal diameter ascertained, in order that correction may be made for capillarity. This correction, however, should be combined with the error of graduation, and form a permanent index error, ascertainable at any time by comparison with an acknowledged standard barometer.

The barometer is supported in the tripod stand (furnished as part of the instrument) when used for observation. It is suspended by placing two studs, in the ring on the frame, in slots formed on the top of the stand, so that it hangs freely and vertically in gimbals. To the metal top of the stand, mahogany legs are hinged. To make the barometer portable, it must be lifted out of the stand, sloped gently until the mercury reaches the top, turning the screw at the bottom meanwhile; then invert and screw until the mercury is made tight. The inverted instrument packs in the stand, the legs being formed to fit round the frame ; and receptacles are scooped out for the cistern, thermometer, gimbals, and vernier; so that the instrument is firmly surrounded by the wooden legs, which are held fast together by brass rings passed over them.

Fig. 29.		**39. Newman's Mountain Barometer.**—Fig. 29 is an illustration of the mountain barometer known as Newman's. The cistern consists of two separate compartments ;—the top of the lower and the bottom of the upper, being perfectly flat, are pivoted closely together at the centres, so that the lower can move through a small arc, when turned by the hand. This movement is limited by two stops. The top of the lower compartment and the bottom of the upper have each a circular hole, through which the mercury communicates. When the instrument is required for observation, the cistern is turned close up to the stop marked "*open*" or "*not portable.*" When it is necessary to pack it for travelling, the mercurial column must be allowed to fill the tube by sloping the barometer gently; then invert it, and move the cistern to the stop marked "*shut*" or "*portable.*" In this condition, the upper compartment is completely filled with mercury, and consequently that in the tube cannot move about, so as to admit air or endanger the tube. Nor can the mercury pass back to the lower compartment, as the holes are not now coincident, and the contact is made too perfect to allow the mercury to creep between the surfaces. The tube does not enter the lower compartment, which is completely full of mercury when the instrument is arranged for observation. The spare capacity of the upper cistern is sufficient to receive the mercury which descends from the tube to the limit of the engraved scale, which in these barometers generally extends only to about 20 inches. A lower limit could of course be given by increasing the size of the cisterns, which it is not advisable to do unless for a special purpose. This barometer may be had mounted in wood, or in brass frame. If in wood, it has a brass shield, which slides round the scale part of the frame, so as to be easily brought in front of the tube and scale as a protection in travelling ; the vernier screw, in this case, being placed at the top of the instrument. When the scale is graduated with true inches, the neutral point, the capacity and capillarity corrections should be marked on the frame. The graduated scales, however, placed on these barometers in brass frames, are usually artificial inches, like the Kew plan of graduation; the advantage being that one simple correction only is required, viz. one for index error and capillarity combined, which can always be readily determined by comparison with a standard barometer; moreover, as no adjustment of cistern is

required in reading, the instrument can be verified by artificial pressure throughout the scale, by the plan practised at Kew, Liverpool, &c., and already described (see p. 18).

40. NEGRETTI & ZAMBRA'S PATENT MOUNTAIN AND OTHER BAROMETERS.

This invention is intended to make mountain and other barometers of standard accuracy stronger, more portable, and less liable to derangement, when being carried about, than heretofore, by dispensing with the ordinary flexible cistern containing the mercury at the bottom of the instrument, and adapting in lieu thereof a rigid cistern constructed of glass and iron. The cistern is composed of a glass cylinder, which is secured in a metallic tube or frame. In order to render the cistern mercury-tight at top and bottom, metal caps are screwed into the tube or frame, and bear against leather washers placed between them and the edges of the glass cylinder. The upper cap of the cistern is tapped with a fine threaded screw to receive the iron plug or socket, into which the barometer tube is securely fixed. The whole length of this plug has a fine screw cut upon it by which the cistern can be screwed up or down. At the side of this plug or socket, extending from the lower end to within half an inch of the top, is cut a groove for admitting the air to the surface of the mercury within the cistern when the barometer is in use. An ivory point is screwed into the under surface of the plug, carrying the barometer tube. This ivory point is very carefully adjusted by measurement to be the zero point of the instrument, from which the barometer scale of inches is divided. The surface of the mercury in the cistern is adjusted to the zero point by screwing the cistern up or down until the ivory point and its reflected image are in contact.

Fig. 30:

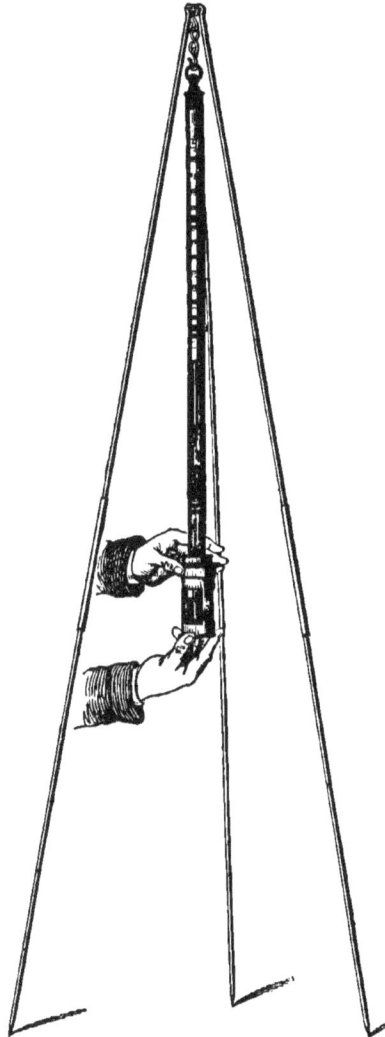

The instrument (fig. 30) is shown in a state of adjustment, ready to take an observation;. but *when it is desired to render it portable, it must be inclined, until mercury from the cistern fills the tube; the cistern must then be screwed up on the socket*, so as to bring the face of the upper cap against the under side of the shoulder of the cover immediately above it; the instrument may then be carried without being liable to derangement.

Precautions necessary in using the Mountain Barometer.—On removing the barometer from its case after a journey, allow it to remain with its scale end downward, whilst the cistern is unscrewed to the extent of *one turn of the screw*, after which slightly shake the cistern ; the mercury in it will then completely fill the end of the barometer tube, should any portion of it have escaped therefrom.

The barometer is then inverted, and if it be desired to make an observation, suspend it vertically from its stand by the ring at top. The cistern must then be unscrewed, until the surface of the mercury is brought just level with the extreme end of the ivory or zero point fixed to the iron plug on which the glass cistern moves up and down.

Should the elevation of the place where the barometer is to be used be considerably above the sea level, it will be well—after suspending it from the stand— to unscrew the cistern several turns, *holding the barometer in an oblique position*, as at great heights the mercury will fall considerably quicker than the cistern can be unscrewed, thereby filling it to overflowing; but by partly unscrewing the cistern first, room is given for the reception of a fall of mercury to the extent of several inches.

The cistern must not be unscrewed when the *Instrument is* INVERTED *more than* two turns of the screw, otherwise the mercury will flow out through the groove.

It is found safer when travelling to carry the barometer in a horizontal position, or with its cistern end uppermost.

To clean the Barometer.—Should at any time the mercury in the cistern become oxidised, and reading from its surface be difficult, it can be readily cleaned by removing the cistern and its contained mercury from the barometer frame by unscrewing it *when in a horizontal position;* this precaution is necessary that the mercury in the tube may not escape, and thereby allow air to enter. The cistern must then be emptied, and with a dry clean leather, or silk handkerchief, well cleaned.

The operation of cleaning being performed, return the cistern to the frame, and screw it until the face is brought up against the under side of the shoulder, still keeping the instrument *horizontal.* The cistern is now ready for re-filling, to do which stand the barometer on end *head downwards*, and remove the small screw at bottom; through the aperture thus opened, pour in mercury, passing it through a paper funnel with a very small aperture. It is well to pass the mercury through a very small funnel two or three times before returning it to the barometer cistern, as by this process all particles of dust or oxide adhere to the paper, and are effectually removed.

Should any small quantity of the mercury be lost during the operation of cleaning, it is of no importance so long as sufficient remains to allow of adjustment to the zero point. This latter constitutes one of the great advantages of this new instrument over the ordinary barometer; for, in the majority of cases, after an instrument has been compared carefully with a standard, should mercury be lost, there is no means of correcting the error unless a standard barometer be at hand; the new barometer is, in this respect, independent, a little mercury more or less being unimportant.

41. Short Tube Barometer.—This is simply a tube shorter, as may be required, than that necessary to show the atmospheric pressure at the sea level. It is convenient for balloon purposes, and for use at mountain stations, being of course a special construction.

42. Method of Calculating Heights by the Barometer.—The pressure of the atmosphere being measured by the barometer, it is evident that as the instrument is carried up a high mountain or elevated in a balloon, the length of the column must decrease as the atmospheric pressure decreases, in consequence of a stratum of air being left below. The pressure of air arises from its weight, or the attraction of gravitation upon it, and therefore the quantity of air below the barometer cistern cannot influence the height of the column. Hence it follows that a certain relation must exist between the difference of the barometric pressure at the foot and at the top of a hill or other elevation, and the difference of the absolute heights above the sea. Theoretical investigation, abundantly confirmed by practical results, has determined that the strata of air decrease in density in a geometrical proportion, while the elevations increase in an arithmetical one. Hence we have a method of determining differences of level, by observations made on the density of the air by means of the barometer. It is beyond our purpose to explain in detail the principles upon which this method is founded, or to give its mathematical investigation. We append Tables, which will be useful to practical persons,—surveyors, engineers, travellers, tourists, &c.,—who may carry a barometer as a travelling companion.

Table I. is calculated from the formula, height in feet $= 60,200$ (log. 29·922 —log. B) $+ 925$; where 29·922 is the mean atmospheric pressure at 32° F., and the mean sea-level in latitude 45°; and B is any other barometric pressure; the 925 being added to avoid minus signs in the Table.

Table II. contains the correction necessary for the mean temperature of the stratum of air between the stations of observation; and is computed from Regnault's co-efficient for the expansion of air, which is ·002036 of its volume at 32° for each degree above that temperature.

Table III. is the correction due to the difference of gravitation in any other latitude, and is found from the formula, $x = 1 + ·00265$ cos. 2 lat.

Table IV. is to correct for the diminution of gravity in ascending from the sea-level.

To use these Tables: The barometer readings at the upper and lower stations having been corrected and reduced to temperature 32° F., take out from Table I. the numbers opposite the corrected readings, and subtract the lower from the upper. Multiply this difference successively by the factors found in Tables II. and III. The factor from Table III. may be neglected unless precision is desired. Finally, add the correction taken from Table IV.

TABLE I.

Approximate Height due to Barometric Pressure.

Inches.	Feet.	Inches.	Feet.	Inches.	Feet.
31·0	0	28·2	2475	25·4	5209
30·9	84	·1	2568	·3	5312
·8	169	28·0	2661	·2	5415
·7	254	27·9	2754	·1	5519
·6	339	·8	2848	25·0	5623
·5	425	·7	2942	24·9	5728
·4	511	·6	3037	·8	5833
·3	597	·5	3132	·7	5939
·2	683	·4	3227	·6	6045
·1	770	·3	3323	·5	6152
30·0	857	·2	3419	·4	6259
29·9	944	·1	3515	·3	6366
·8	1032	27·0	3612	·2	6474
·7	1120	26·9	3709	·1	6582
·6	1208	·8	3806	24·0	6691
·5	1296	·7	3904	23·9	6800
·4	1385	·6	4002	·8	6910
·3	1474	·5	4100	·7	7020
·2	1563	·4	4199	·6	7131
·1	1653	·3	4298	·5	7242
29·0	1743	·2	4398	·4	7353
28·9	1833	·1	4498	·3	7465
·8	1924	26·0	4598	·2	7577
·7	2015	25·9	4699	·1	7690
·6	2106	·8	4800	23·0	7803
·5	2198	·7	4902	22·9	7917
·4	2290	·6	5004	·8	8032
·3	2382	·5	5106	·7	8147

TABLE I.—*continued.*

Approximate Height due to Barometric Pressure.

Inches.	Feet.	Inches.	Feet.	Inches.	Feet.
22·6	8262	18·9	12937	15·2	18632
·5	8378	·8	13076	·1	18805
·4	8495	·7	13215	15·0	18979
·3	8612	·6	13355	14·9	19154
·2	8729	·5	13496	·8	19330
·1	8847	·4	13638	·7	19507
22·0	8966	·3	13780	·6	19685
21·9	9085	·2	13923	·5	19865
·8	9205	·1	14067	·4	20046
·7	9325	18·0	14212	·3	20228
·6	9446	17·9	14358	·2	20412
·5	9567	·8	14505	·1	20597
·4	9689	·7	14652	14·0	20788
·3	9811	·6	14800	13·9	20970
·2	9934	·5	14949	·8	21159
·1	10058	·4	15099	·7	21349
21·0	10182	·3	15250	·6	21541
20·9	10307	·2	15402	·5	21734
·8	10432	·1	15554	·4	21928
·7	10558	17·0	15707	·3	22124
·6	10684	16·9	15861	·2	22321
·5	10812	·8	16016	·1	22520
·4	10940	·7	16172	13·0	22720
·3	11069	·6	16329	12·9	22922
·2	11198	·5	16487	·8	23126
·1	11328	·4	16646	·7	23331
20·0	11458	·3	16806	·6	23538
19·9	11589	·2	16967	·5	23746
·8	11721	·1	17129	·4	23956
·7	11853	16·0	17292	·3	24168
·6	11986	15·9	17456	·2	24381
·5	12120	·8	17621	·1	24596
·4	12254	·7	17787	12·0	24813
·3	12389	·6	17954	11·9	25032
·2	12525	·5	18122	·8	25253
·1	12662	·4	18291	·7	25476
19·0	12799	·3	18461	·6	25700

TABLE II.

Correction due to Mean Temperature of the Air.

Mean Temp.	Factor.	Mean Temp.	Factor.	Mean Temp.	Factor.
10°	0·955	85°	1·006	60°	1·057
11	·957	86	1·008	61	1·059
12	·959	87	1·010	62	1·061
18	·961	88	1·012	68	1·063
14	·963	89	1·014	64	1·065
15	·965	40	1·016	65	1·067
16	·967	41	1·018	66	1·069
17	·969	42	1·020	67	1·071
18	·971	48	1·022	68	1·078
19	·974	44	1·024	69	1·075
20	·976	45	1·026	70	1·077
21	·978	46	1·029	71	1·079
22	·980	47	1·031	72	1·081
28	·982	48	1·033	78	1·083
24	·984	49	1·035	74	1·086
25	·986	50	1·037	75	1·088
26	·988	51	1·039	76	1·090
27	.990	52	1·041	77	1·092
28	·992	58	1·048	78	1·094
29	·994	54	1·045	79	1·096
80	·996	55	1.047	80	1·098
81	0·998	56	1·049	81	1·100
82	1·000	57	1·051	82	1·102
88	1·002	58	1·053	88	1·104
34	1·004	59	1·055	84	1·106

TABLE III.

Latitude.	Factor.	Latitude.	Factor.	Latitude.	Factor.
80°	0·99751	50	0·99954	20	1·00203
75	0·99770	45	1·00000	15	1·00230
70	0·99797	40	1·00046	10	1·00249
65	0·99830	85	1·00090	5	1·00261
60	0·99868	80	1·00132	0	1·00265
55	0·99910	25	1·00170		

TABLE IV.

Height in Thousand Feet.	Correction Additive.	Height in Thousand Feet.	Correction Additive.
1	3	14	44
2	5	15	48
3	8	16	52
4	11	17	56
5	14	18	60
6	17	19	65
7	20	20	69
8	23	21	74
9	26	22	78
10	30	23 ·	83
11	33	24	88
12	37	25	93
13	41	26	98

EXAMPLE 1. On October 21st, 1852, when Mr. Welsh ascended in a balloon, at 3h. 30m. p.m., the barometer, corrected and reduced, was 18·85, the air temperature 27°, while at Greenwich, 159 feet above the sea, the barometer at the same time was 29·97 inches, air temperature 49°, the balloon not being more than 5 miles S.W. from over Greenwich; required its elevation.

				Feet.
Barometer · in Balloon	...	18·85, Table I.	=	13007
,, at Greenwich	...	29·97	,,	883
				12124
Mean Temperature, 38°, Table II. Factor	...			1·012
				12269·
Latitude 51¼°, Factor from Table III.				·99941
				12262
Correction from Table IV.	...	...		38
				12300
Elevation of Greenwich	...	...		159
,, Balloon				12459 feet.

The following examples, from the balloon ascents of J. Glashier, Esq., F.R.S., will serve for practice.*

---

* *Vide* Report of the British Association, 1862.

2.   Ascended from Wolverhampton, 18th August, 1862, at 2h. 38m. p.m.; barometer (in all cases corrected and reduced to 32° F) was 14·868, the temperature of the air 26°; at the same time, at Wrottesley Hall, 581 feet above the sea, in latitude 52¼° N, the barometer was 29·46, and the temperature of the air 65°·4; find the elevation of the balloon above the sea.

Height, 18,959 feet.

3.   From the same place an ascent was made 5th September, 1862, when at 1h. 48m. p.m. barometer was 11·954, air 0°; at Wrottesley Hall 29·38, air 56°.

Height, 23,923 feet.

4.   From the Crystal Palace a balloon ascent was made 20th August, 1862. At 6h. 47m. p.m. barometer was 25·55, air 50°·5; and at the same time at Greenwich Observatory, at 159 feet above the sea, the barometer was 29·81, air 63°.

Height, 4,406 feet.

5.   From the same place an ascent was made 8th September, 1862. At 5 p.m., the balloon being over Blackheath, barometer was 25·60, and the air 49°·5, while at Greenwich, barometer was 29·92, air 66°·4.

Height, 4,461 feet.

CHAPTER V.

SECONDARY BAROMETERS.

43. Desirability of Magnifying the Barometer Range.—The limits within which the ordinary barometric column oscillates, do not exceed four inches for extreme range, while the ordinary range is confined to about two inches; hence it has often been felt that the public utility of the instrument would be greatly enhanced if by any means the scale indications could be increased in length. This object was sought to be obtained by bending the upper part of the tube from the vertical, so that the inches on the scale could be increased in length in proportion to the secant of the angle it made with the vertical. This was called "the diagonal barometer." The upper part of the tube has also been formed into a spiral, and the scale, placed along it, is thus greatly enlarged.

But these methods of enlarging the indications cannot be so successfully accomplished, nor so cheaply nor so elegantly, as is done by the principle employed in the dial barometer. Hence they are not in use.

44. Howson's Long Range Barometer.—Very recently quite a novel design has been patented by Mr. Howson, for a long range barometer. The construction requires neither distortion of the tube, nor mechanism for converting a short scale into a long one; but the mercury itself rises and falls, through an extended range, naturally, and in simple obedience to the varying pressure of the atmosphere. The tube is fixed, but its cistern is sustained by the mere pressure of the atmosphere. Looking at the instrument, it seems a perfect marvel. It appears as though the cistern with the mercury in it must fall to the ground. The bore of the tube is wide, about an inch across. A long glass rod is fixed to the bottom of the glass cistern, where a piece of cork or some elastic substance is also placed. The tube is filled with mercury; the glass rod is plunged into the tube as it is held top downwards, until the cork gets close up to the tube and fits tightly against it. The pressure against the cork simply prevents the mercury from coming out while the instrument is being inverted. When it is inverted, the mercury partly falls, and forms an ordinary barometric column. When the top is held, the cistern and glass rod, instead of falling away, remain perfectly suspended. There is no material support to the cistern; the tube only is fixed, the cistern hangs to it. Glass is many times lighter than mercury. When the glass rod is introduced, it displaces an equal volume of mercury. The glass rod, being so much lighter than mercury, floats and sustains the additional weight of the cistern by its buoyancy. In the mean time, the atmosphere is acting upon the mercury, keeping up the ordinary barometric column. Supposing there is a rise in the ordinary barometer, the

Fig. 31.

atmosphere presses some more mercury up the tube. This mercury is taken out of the cistern, which of course becomes lighter, and therefore the rod and cistern float up a little higher, which thus causes the column of mercury to rise still more. The increased pressure and buoyancy thus acting together, increase the ascent in the barometric column, as shown by the fixed scale. One inch in the barometer might be represented by two or more inches in this instrument, according to construction. Supposing there was a decrease of pressure, the mercury would fall, come into the cistern, make it heavier, and increase the fall somewhat. Friction guides, at the top of the rod, prevent it coming into contact with the side of the tube when vertically suspended. The illustration, Fig. 31, shows the appearance of the instrument as framed in wood by the makers, Messrs. Negretti and Zambra.

45. McNeild's Long Range Barometer.—A barometer designed by a gentleman named McNeild is on a directly opposite principle to the one just described. The tube is made to float on the mercury in the cistern. It is filled with mercury, inverted in the usual manner, then allowed to float, being held vertically by glass friction points or guides. By this contrivance, the ordinary range of the barometer is greatly increased. One inch rise or fall in the standard barometer may be represented by four or five inches in this instrument, so that it shows small variations in atmospheric pressure very distinctly. As the mercury falls in the tube with a decrease of pressure, the surface of the mercury in the cistern rises, and the floating tube rises also, which causes an additional descent in the column, as shown by fixed graduations on the tube. With an increase of pressure, some mercury will leave the cistern and rise in the tube, while the tube itself will fall, and so cause an additional ascent of mercury. This barometer is identical in principle with King's Barograph (see p. 34).

The construction of Howson's and McNeild's Barometers has been assigned to Messrs. Negretti and Zambra. These instruments are usually made for domestic purposes with a scale of from three to five, and for public use from five to eight times the scale of the ordinary standard. Their sensitiveness is consequently increased in an equal proportion, and they have the additional advantage of not being affected by differences of level in the cistern. However, these novelties have not been sufficiently tried to determine their practical value for strictly scientific purposes; but as weather-glasses, for showing minute changes, they are superior to the common barometer.

46. The Water-glass Barometer.—If a Florence flask, having a long neck, have a small quantity of water poured into it, and then be inverted and so supported that the open end dips into a vessel containing water, a small column of water will be confined in the neck of the bottle, the pressure of which, upon the surface of the exposed water, will be equal to the difference between the atmospheric pressure and the elasticity of the confined air in the body of the bottle. As the pressure of the atmosphere varies, this column will alter in height. But the elasticity of the confined air is also subject to variations, owing to changes of temperature. It follows, then, that the oscillations of the column are dependent on alterations of temperature and atmospheric pressure. Such an arrangement has been called "the Water-glass Barometer," and bears about the same relative value to the mercurial barometer, as an exponent of weather changes, that a cat-gut hygrometer bears to a thermometric hygrometer, as an indicator of relative moisture.

47. SYMPIESOMETER.

Nevertheless the instrument now about to be described, depending upon similar principles, but scientifically constructed and graduated, is a very useful and valuable substitute for the mercurial barometer. It consists of a glass tube, varying, according

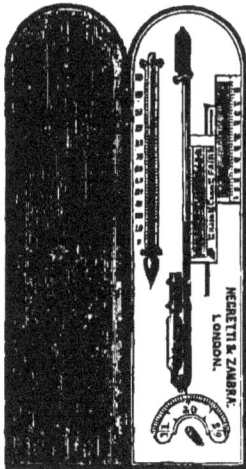

Fig. 32.

to the purposes for which the instrument is required, from six to twenty-four inches in length. The upper end is closed, and formed into a bulb; the lower is turned up, formed into a cistern, and open at top, through a pipette, or cone. A plug, moveable by a catch from below, can be made to close this opening, so as to render the instrument portable.

The upper portion of the tube is filled with air; the lower portion, and part of the cistern, with sulphuric acid, coloured so as to render it plainly visible. Formerly, hydrogen and oil were used. It was found, however, that, by the process known to chemists as *osmosis*, this light gas in time partially escaped, and the remainder became mixed with air, the consequence being that the graduations were no longer correct. They are more durable as at present constructed. The liquid rises and falls in the tube with the variations of atmospheric pressure and temperature acting together. If the pressure were constant, the confined air would expand and contract for temperature only, and the instrument would act as a thermometer. In fact, the instrument is regarded as such in the manufacture; and the thermometric scales are ascertained and engraved on the scale. A good mercurial thermometer is also mounted on the same frame. If, therefore, at any time the mercurial and the air thermometers do not read alike, it must evidently be due to the atmospheric pressure acting upon the air in the tube; and it is further evident that, under these circumstances, the position of the top of the liquid may be marked to represent the barometric pressure at the time. In this manner a scale of pressure is ascertained by comparison with a standard barometer, extending generally from 27 to 31 inches.

When made correctly, these instruments agree well with the mercurial barometer for a number of years, and their subsequent adjustment is not a matter of much expense.

For use at sea, the liquid column is contracted at the bend. The sympiesometer is very sensitive, and feels the alterations in the atmospheric pressure sooner than the ordinary marine barometer.

The scale is usually on silvered brass, mounted on a mahogany or rosewood frame, protected in front by plate glass. It is generally furnished with a revolving register, to record the observation, in order that it may be known whether the pressure has increased or decreased in the interval of observation.

Small pocket sympiesometers are sometimes fitted with ivory scales, and protected by a neat velvet-lined pasteboard or morocco case.

How to take an Observation.—In practice, the indications of the atmospheric pressure are obtained from the sympiesometer by noting, first, the temperature of the mercurial thermometer; secondly, adjusting the pointer of the pressure scale to the same degree of temperature on the scale of the air column; thirdly, reading the height of the liquid on the sliding scale.

Directions for Use.—The sympiesometer should be carried and handled so as to keep the top always upwards, to prevent the air mechanically mixing with the liquid. Care should also be taken to screen it from casual rays of the sun or cabin fire.

48. ANEROIDS.

The beautiful and highly ingenious instrument called by the name *Aneroid*, is no less remarkable for the scientific principles of its construction and action, than for the nicety of its mechanism. It is a substitute, and perhaps the best of all substitutes, for the mercurial barometer. As its name implies, it is constructed "without fluid." It was invented by M. Vidi of Paris. In the general form in which it is made it consists of a brass cylindrical case about four inches in diameter and one and a half inch deep, faced with a dial graduated and marked similarly to the dial-plate of a "wheel-barometer," upon which the index or pointer shows the atmospheric pressure in inches and decimals of an inch in accordance with the mercurial barometer. Within the case, for ordinary sizes, is placed a flat metal box, generally not more than half an inch thick and about two inches or a little more in diameter, from which nearly all the air is exhausted. The top and bottom of this box is corrugated in concentric circles, so as to yield inwardly to external pressure, and return when the pressure is removed. The pressure of the atmosphere, acting externally, continually changes, while the elastic pressure of the small quantity of air within can only vary by its volume being increased or decreased, or by change of temperature. Leaving out of consideration, for the moment, the effect of temperature, we can readily perceive that as the pressure is lessened upon the outside of the box, the elastic force of the air within will force out the top and bottom of the box; and when the outer pressure is increased they will be forced in. Thus with the varying pressure of the atmosphere, the top and bottom of the box approach to and recede from each other by a small quantity; but the bottom being fixed, nearly all this motion takes place on the top. Thus the top of the box is like an elastic cushion, which rises and falls according as the compressing force lessens or increases. To the eye these expansions and contractions would not be perceptible, so small is the motion. But they are rendered very evident by a nice mechanical arrangement. To the box is attached a strong piece of iron, kept pressed upon it by a spring at one extremity; so that as the top of the box rises, the motion is made sensible at the point held by the spring, and when the top descends the spring draws the piece of iron into close contact with it. This piece of iron acts as a lever, having its fulcrum at one extremity, the power at the centre of the box-top, and the other extremity controlled by the spring. Thus it is evident that the small motion of the centre of the box-top is much increased at the spring extremity. The motion thus obtained is communicated to a system of levers; and, by the intervention of a piece of watch-chain and a fine spring passing round the arbour, turns the index to the right or left, according as the external pressure increases or decreases. Thus, when by increase of pressure the vacuum box is compressed, the mechanism transfers the movement to the index, and it moves to the right; when the vacuum box bulges out under diminished pressure, the mechanical motion is reversed, and the index moves to the left. As the index traverses the dial, it shows upon the scale the pressure corresponding with that which a good mercurial barometer would at the same time and place indicate; that is, supposing it correctly adjusted.

A different and more elegant arrangement has since been adopted. A broad curved spring is connected to the top of the vacuum box, so as to be compressed by the top of the box yielding inward to increased pressure, and to relax itself and the box as the pressure is lessened. The system of levers is connected to this spring, which augments and transfers the motion to the index, in the manner already described. Increase of pressure causes the levers to slacken the piece of watch-chain connected with them and the arbour of the index. The spring now uncoils, winds the chain upon the arbour, and turns the index to the right. Decrease of pressure winds the chain off the barrel, tightens the spiral spring, which thus turns the index to the left. The graduations of the aneroid scale are obtained by comparisons with the correct standard reading of a mercurial barometer, under the normal and reduced atmospheric pressure. Reduced pressure is obtained by placing both instruments under the receiver of an air pump.

Fig. 33 represents the latest improved mechanism of an aneroid. The outer case and the face of the instrument are removed, but the hand is attached by its collet to the arbour. A is the corrugated box, which has been exhausted of air through the tube, J, and hermetically sealed by solder-

Fig. 33.

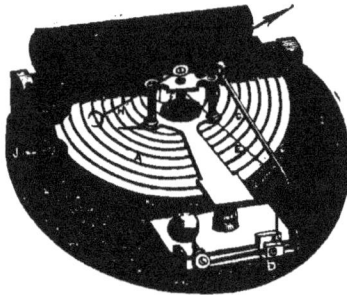

ing. B is a powerful curved spring, resting in gudgeons fixed on the frame-plate, and attached to a socket behind, F, in the top of the box. A lever, C, joined to the stout edge of the spring, is connected, by the bent lever at D, with the chain, E, the other end of which is coiled round, and fastened to the arbour, F. As the box, A, is compressed by the weight of the atmosphere increasing, the spring, B, is tightened, the lever, C, depressed, and the chain, E, uncoiled from F, which is thereby turned so that the hand, H, moves to the right. In the mean while the spiral spring, G, coiled round F, and fixed at one extremity to the frame-work and by the other to F, is compressed. When, therefore, the pressure decreases, A and B relax, by virtue of their elasticity; E slackens, G unwinds, turning F, which carries H to the left. Near J is shown an iron pillar, cast as part of the stock of the spring, B. A screw works in this pillar through the bottom of the plate, by means of which the spring, B, may be so adjusted to the box, A, as to set the hand, H, to read on the scale according to the indications of a mercurial barometer. The lever, C, is composed of brass and steel, soldered together, and adjusted by repeated trials to correct for the effects of temperature.

A thermometer is sometimes attached to the aneroid, as it is convenient for indicating the temperature of the air. As regards the instrument itself, no correction for temperature can be applied with certainty. It should be set to read with the mercurial barometer at 32° F. Then the readings from it are supposed to require no correction.

In considering the effects of temperature upon the aneroid, they are found to be somewhat complex. There is the effect of expansion and contraction of the various metals of which the mechanism is composed; and there is the effect on the elasticity of the small portion of air in the box. An increase of temperature produces greater, a diminution less elasticity in this air. The compensation for effects of temperature is adjusted by the process of "trial and error," and only a few makers do it well. It is very often a mere sham. Admiral FitzRoy writes, in his *Barometer Manual*, "The known expansion and contraction of metals under varying temperatures, caused doubts as to the accuracy of the aneroid under such changes; but they were partly removed by introducing into the vacuum box a small portion of gas, as a compensation for the effects of heat or cold. The gas in the box, changing i's bulk on a change of temperature, was intended to compensate for the effect on the metals of which the aneroid is made. Besides which, a further and more reliable compensation has lately been effected by a combination of brass and steel bars."

"Aneroid barometers, if often compared with good mercurial columns, are similar in their indications, and valuable; but it must be remembered that they are not independent instruments, that they are set originally by a barometer, require adjustment occasionally, and may deteriorate in time, though slowly."

"The aneroid is quick in showing the variation of atmospheric pressure; and to the navigator who knows the difficulty, at times, of using barometers, this instrument is a great boon, for it can be placed anywhere, quite out of harm's way, and is not affected by the ship's motion, although faithfully giving indication of increased or diminished pressure of air. In ascending or descending elevations, the hand of the aneroid may be seen to move (like the hand of a watch), showing the height above the level of the sea, or the difference of level between places of comparison."

In the admiral's *Notes on Meteorology*, he says, "The aneroid is an excellent *weather glass*, if well made. Compensation for heat or cold has lately been introduced by efficient mechanism. In its *improved* condition, when the cost may be about £5, it is fit for measuring heights as far as 5,000 feet with approximate accuracy; but even at the price of £3, as a *weather-glass* only, it is exceedingly valuable, because it can be carried anywhere; and if now and then compared with a good barometer, it may be relied on sufficiently. I have had one in constant use for ten years, and it appears to be as good now as at first. For a ship of war (considering concussion by the fire of guns), for boats, or to put in a drawer, or on a table, I believe there is nothing better than it for use as a common weather-glass."

Colonel Sir H. James, R.E., in his *Instructions for taking Meteorological Observations*, says of the aneroid, "This is a most valuable instrument; it is extremely portable. I have had one in use for upwards of ten years, and find it to be the best form of barometer, as a "weather-glass," that has been made."

One of the objects of Mr. Glaisher's experiments in balloons was "to compare the readings of an aneroid barometer with those of a mercurial barometer up to five miles." In the comparisons the readings of the mercurial barometer were corrected for index-error and temperature. The aneroid readings, says Mr. Glaisher, "prove all the observations made in the several ascents may be safely depended upon, and also that an aneroid barometer can be made to read correctly to pressures below twelve inches." As one of the general conclusions derived from his experiments he states, "that an aneroid barometer read correctly to the first place, and probably to the second place of decimals, to a pressure as low as seven inches." The two aneroids used by Mr. Glaisher were by Messrs. Negretti and Zambra.

Aneroids are now manufactured almost perfectly compensated for temperature. Such an instrument therefore ought to show the same pressure in the external air at a temperature say of 40°, as it would in a room where the temperature at the same time may be 60°; provided there is no difference of elevation. To test it thoroughly would require an examination and a comparison with barometer readings reduced to 32° F., conducted through a long range of temperature and under artificially reduced pressure. A practical method appears to be to compare the aneroid daily, or more often, for a few weeks with the readings of a mercurial barometer reduced to 32°; and if the error so found be constant, the object of the compensation may be assumed to be attained, particularly if the temperature during the period has varied greatly.

Directions for using the Aneroid.—Aneroids are generally suspended with the dial vertical; but if they be placed with the dial horizontal, the indications differ a few hundredths of an inch in the two positions. Hence, if their indications are registered, they should be kept in the same position.

The aneroid will not answer for exact scientific purposes, as it cannot be relied upon for a length of time. Its error of indication changes slowly, and hence the necessity of its being set from time to time with the reading of a good barometer. To allow of this being done, at the back of the outer case is the head of a screw in connection with the spring attached to the vacuum box. By applying a small turnscrew to this screw, the spring of the vacuum box may be tightened or relaxed, and the index made to move correspondingly to the right or left on the dial. By this means, besides being enabled to correct the aneroid at any time, " if the measure of a height rather greater than the aneroid will commonly show be required, it may be *re-set* thus : When at the upper station (*within its range*), and having noted the reading carefully, touch the screw behind so as to bring back the hand a few inches (if the instrument will admit), then read off and start again. *Reverse the operation when descending.* This may add some inches of measure *approximately.*"—*FitzRoy.*

49. Small Size Aneroids.—The patent for the Aneroid having expired, Admiral FitzRoy urged upon Messrs. Negretti & Zambra the desirability of reducing the size at which it had hitherto been made, as well as of improving its mechanical arrangement, and compensation for temperature. They accordingly engaged skilful workmen, who, under their directions, and at their expense, by a great

Fig. 34.

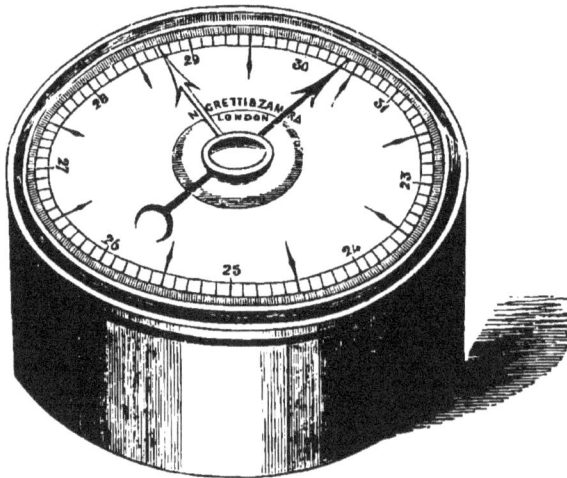

amount of labour and experiment, succeeded in reducing its dimensions to two inches in diameter, and an inch and a quarter thick. The exact size and appearance of this aneroid are shown in fig. 34. The compensation is carefully adjusted, and the graduations of the dial ascertained under reduced pressure, so that they are not quite equal, but more accurate.

50. Watch Aneroid.—Subsequently the aneroid has been further reduced in size and it can now be had from an inch and a quarter to six inches in diameter. The smallest size can be enclosed in watch cases, fig. 35, or otherwise, so as to be adapted to the pocket. By a beautifully simple contrivance, a milled rim is adjusted to move round with hand pressure, and carry a fine index or pointer, outside and around the scale engraved on the dial, or face, for the purpose of marking the reading, so that the subsequent increase or decrease of pressure may be readily seen. These very small instruments are found to act quite as correctly as the largest, and are much more serviceable. Besides serving the purpose of a weather-glass in the house or away from home, if carried in the pocket, they are admirably suited to the exigencies of tourists and travellers. They may be had with scale sufficient to measure heights not exceeding 8,000 feet; with a scale of elevation in feet, as well as of pressure in inches, engraved on the dial. The scale of elevation, which is for the temperature of 50°, was computed by Professor Airy, the Astronomer Royal, who kindly presented it to Messrs. Negretti and Zambra, at the same time suggesting its application. Moderate-sized aneroids, fitted in leathern sling cases, are also good travelling instruments, and will be found serviceable to pilots, fishermen, and for use in coasting and small vessels, where a mercurial barometer cannot be employed, because requiring too much space.

Admiral FitzRoy, in a communication to the *Mercantile Marine Magazine*, December, 1860, says :—" Aneroids are now made more portable, so that a pilot or chief boatman may carry one in his pocket, as a railway guard carries his timekeeper ; and, thus provided, pilots cruising for expected ships would be able to caution strangers arriving, if bad weather were impending, or give warning to coasters or fishing boats. Harbours of Refuge, however excellent and important, are not always accessible, even when most wanted, as in snow, rain, or darkness, when neither land, nor buoy, nor even a lighthouse-light can be seen."

Fig. 35.

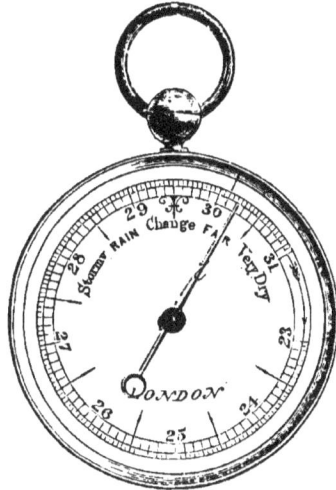

51. Measurement of Heights by the Aneroid.—For measuring heights not exceeding many hundred feet above the sea-level by means of the aneroid, the following simple method will suffice:—

Divide the difference between the aneroid readings at the lower and upper stations by ·0011; the quotient will give the approximate height in feet.

Thus, supposing the aneroid to read at the

Lower Station .... 30·385 inches.

Upper Station .... 30·025

Difference...... ·360

Divided gives $\dfrac{·360}{·0011} = 827$ feet.

As an illustration of the mode in which the aneroid should be used in measuring heights, the following example is given:—

A gentleman who ascended Helvellyn, August 12th, 1862, recorded the following observations with a pocket aneroid by Negretti and Zambra:—

Near 10 a.m., at the first milestone from Ambleside, found by survey to be 188 feet above the sea, the aneroid read 29·89 inches; about 1 p.m., at the summit of Helvellyn, 26·81; and at 5 p.m., at the milestone again, 29·76. The temperature of the lower air was 57°, of the upper, 54°. Hence the height of the mountain is deduced as follows:—

		Inches.		
Reading at 10 a.m.	...	29·89		
,, 5 p.m.	...	29·76		
Mean	...	29·825	Table I.*	1010
Upper Reading	...	26·81	,,	8796
Difference	... ...			2786
Mean Temperature 55°·5, gives in Table II.	... ...			1·048
				2920
Lat. 55° N., gives in Table III.	...			·9991
				2917
Table IV.				5
Difference of height ...	... ...			2922
Height of lower station	... ...	... ...		188
,, Helvellyn ...				3110
In Sir J. Herschell's *Physical Geography* it is given as	...			3115 ft.

---

* See page 42 for the Tables.

So near an agreement is attributable to the excellence of the aneroid, and the careful accuracy of the observer.

52. METALLIC BAROMETER.

This instrument, the invention of M. Bourdon, has a great resemblance to the aneroid, but is much simpler in arrangement. The inventor has applied the same principle to the construction of metallic steam-pressure gauges. We are here, however, only concerned with it as constructed to indicate atmospheric pressure. It consists of a long slender flattened metallic tube, partially exhausted of air, and hermetically closed at each end, then fixed upon its centre, and bent round so as to make the ends face each other. The transverse section of this tube is an elongated ellipse. The principle of action is this: interior pressure tends to straighten the tube, external pressure causes it to coil more. Hence as the atmospheric pressure decreases, the ends of the tube become more apart.

This movement is augmented and transferred by a mechanical arrangement of small metallic levers to a radius bar, which carries a rack formed on the arc of its circle. This moves a pinion, upon the arbour of which a light pointer, or "hand," is poised, which indicates the pressure upon a dial. When the pressure increases, the ends of the tube approach each other, and the pointer moves from left to right over the dial. The whole mechanism is fixed in a brass case, having a hole at the back for adjusting the instrument to the mercurial barometer by means of a key, which sets the pointer without affecting the levers. The dial is generally open to show the mechanism, and is protected by a glass, to which is fitted a moveable index.

This barometer is very sensitive, and has the advantage of occupying little space, although it has not yet been made so small as the aneroid. Both these instruments admit of a great variety of mounts to render them ornamental. The metallic barometer can be constructed with a small clock in its centre, so as to form a novel and beautiful drawing-room ornament.

Admiral FitzRoy writes, "Metallic barometers, by Bourdon, have not yet been tested in very moist, hot, or cold air for a sufficient time. They are dependent, or secondary instruments, and liable to deterioration. For limited employment, when sufficiently compared, they may be very useful, especially in a few cases of electrical changes, *not foretold or shown by mercury*, which these seem to indicate remarkably."

They are not so well adapted for travellers, nor for measurements of considerable elevations, as aneroids.

CHAPTER VI.

INSTRUMENTS FOR ASCERTAINING TEMPERATURE.

53. Temperature is the energy with which heat affects our sensation of feeling.

Bodies are said to possess the same temperature, when the amounts of heat which they respectively contain act outwardly with the same intensity of transfer or absorption, producing in the one case the sensation of warmth, in the other that of coldness. Instruments used for the determination and estimation of temperatures are called *Thermometers*.

Experience proves that the same body always occupies the same space at the same temperature; and that for every increase or decrease of its temperature, it undergoes a definite dilatation or contraction of its volume. Provided, then, a body suffers no loss of substance or peculiar change of its constituent elements or atoms, while manifesting changes of temperature it will likewise exhibit alterations in volume; the latter may, therefore, be taken as exponents of the former. The expansion and contraction of bodies are adopted as arbitrary measures of changes of temperature; and any substance will serve for a thermometer in which these changes of volume are sensible, and can be rendered measureable.

54. Thermometric Substances.—Thermometers for meteorological and domestic purposes are constructed with liquids, and generally either mercury or alcohol, because their alterations of volume for the same change of temperature are greater than those of solids; while being more manageable, they are preferred to gases. Mercury is of all substances the best adapted for thermometric purposes, as it maintains the liquid state through a great alteration of heat, has a more equable co-efficient of expansion than any other fluid, and is peculiarly sensitive to changes of temperature. The temperature of solidification of mercury, according to Fahrenheit's scale of temperature, is — 40°; and its temperature of ebullition is about 600°. Sulphuric ether, nitric acid, oil of sassafras, and other limpid fluids, have been employed for thermometers.

55. Description of the Thermometer.—The ordinary thermometer consists of a glass tube of very fine bore, having a bulb of thin glass at one extremity, and closed at the other. The bulb and part of the tube contains mercury; the rest of the tube is a vacuum, and affords space for the expansion of the liquid. This arrangement renders very perceptible the alterations in volume of the mercury due to changes of temperature. It is true, the glass expands and contracts also; but only by about one-twentieth of the extent of the mercury. Regarding the bulb, then, as unalterable in size, all the changes in the bulk of the fluid must take place in the tube, and be exhibited by the expansion and contraction of the column, which variations are made to measure changes of temperature.

56. STANDARD THERMOMETER.

Fig 36.

The peculiarities in the construction of thermometers will be best understood by describing the manufacture of a *Standard Thermometer*, which is one of the most accurate make, and the scale of which is divided independently of any comparison with another thermometer. Fig. 36 is an illustration of such an instrument, on a silvered brass scale.

Selection of Tube.—In selecting the glass tube, much care is requisite to ascertain that its bore is perfectly uniform throughout. As received from the glass-house, the tubes are generally, in their interior, portions of very elongated cones, so that the bore is wider at one end than at the other. With due care, however, a proper length of tube can be selected, in which there is no appreciable difference of bore. This is ascertained by introducing into the tube a length of mercury of about a half or a third of an inch, and accurately measuring it in various positions in the tube. To accomplish this, the workman blows a bulb at one end of the tube, and heats the bulb a little to drive out some of the air. Then, placing the open end in mercury, upon cooling the elasticity of the enclosed air diminishes, and the superior pressure of the atmosphere drives in some mercury. The workman stops the process so soon as he judges sufficient mercury has entered. By cooling or heating the bulb, as necessary, the mercury is made to pass from one end of the tube to the other. Should the length of this portion of mercury alter in various parts of the bore, the tube must be rejected. If it is, as nearly as possible, one uniform length, the tube is set aside for filling.

The *bulb* is never blown by the breath, but by an elastic caoutchouc ball containing air, so that the introduction of moisture is avoided. The spherical form is to be preferred; for it is best adapted to resist the varying pressure of the atmosphere. The bulbs should not be too large, or the mercury will take some time to indicate sudden changes of temperature. Cylindrical bulbs are sometimes desirable, as they offer larger surfaces to the mercury, and enable thermometers to be made more sensitive.

The *mercury*, with which the bulb is to be filled, should be quite pure, and freed from moisture and air by recent boiling.

Filling the Tube.—The filling is effected by heating the bulb with the flame of a spirit-lamp, while the open end is embedded in mercury. Upon allowing the bulb to cool, the atmospheric pressure drives some mercury into it; and the process of heating and cooling is thus continued until sufficient mercury is introduced. The mercury is next boiled in the tube, to expel any air or moisture that may be present. In order to close the tube and exclude all air, the artist ascertains that the tube contains the requisite quantity of mercury; then, by holding the bulb over the spirit flame, he causes the mercury to fill the whole of the tube, and dexterously removing it from the source of heat, he, at the same instant, closes it with the flame of a blow-pipe. If any air remain in the tube, it is easily detected; for if the instrument be inverted, the mercury will fall to the extremity of the tube, if there is a perfect vacuum, unless the tube be so finely capillary that its attraction for the mercury is sufficient to overcome the force of gravity, in which case the mercury will retain its position in every situation of the instrument. If, however, the mercury fall and does not reach quite to the extremity of the bore, some air is present, which must be removed.

The Graduation.—The thermometer is now prepared for graduation, the first part of which process is the determination of two fixed points. These are given by the temperatures of melting ice and of the vapour of boiling water. Melting ice has always the same temperature in every place and under all circumstances; provided only that the water from which the ice is congealed is free from salts. The temperature of the vapour of boiling water depends upon the pressure of the atmosphere, but is always constant for the same pressure.

The fixed point corresponding to the temperature of melting ice is called the *freezing point.* It is obtained by keeping the bulb and the part of the tube occupied by mercury immersed in melting ice, until the mercury contracts to a certain point, where it remains stationary. This position of the end of the mercury is then marked upon the tube.

The *boiling point* is not so easily determined, for the barometer must be consulted about the same time. The boiling apparatus is generally constructed of copper. It consists of a cylindrical boiler, heated from the base by a spirit lamp or charcoal fire. An open tube two or three inches in diameter and of suitable length enters the top of the boiler. This tube is enveloped by another fixed to the top of the boiler but not opening into it, and so that the two tubes are about an inch apart. The object of the outer tube is to protect the inner tube from the cold temperature of the air. The outer tube has an opening at the top for the admission of the thermometer, and a hole near the bottom for the escape of steam through a spout. When the water is made to boil, the steam rises in the inner tube, fills the space between the tubes, and escapes at the spout. The thermometer is then passed down into the inner cylinder, and held securely from the top by means of a piece of caoutchouc. The tubes or cylinders should be of sufficient length to prevent the thermometer entering the water. This is necessary because the temperature of boiling water is influenced by any substance which it holds in chemical solution; and, moreover, its temperature increases with the depth, owing to the pressure of the upper stratum. The thermometer being thus surrounded with steam, the mercury rises in the tube. As it does so, the tube should be depressed so as always to keep the top of the mercury just perceptible. When the temperature of the vapour is attained, the mercury ceases to rise, and remains stationary. The position of the end of the mercury is now marked upon the tube, and the *"boiling-point"* is obtained.

57. Methods of ascertaining the exact Boiling Temperature.—The normal boiling temperature of water all nations have tacitly agreed to fix under a normal barometric pressure of 29·922 inches of mercury, having the temperature of melting ice, in the latitude of 45°, and at the sea-level. If the atmospheric pressure at the time or place of graduating a thermometer does not equal this, the boiling temperature will be higher or lower according as the pressure is greater or less. Hence a reading must be taken from a reliable barometer, which must also be corrected for errors and temperature, and reduced for latitude, in order to compare the actual atmospheric pressure at the time with the assumed normal pressure. Tables of vapour tension, as they are termed, have been computed from accurate experimental investigations and theory,—giving the temperatures of the vapour of water for all probable pressures ; Regnault's, the most recent, is considered the most accurate; and his investigations are based upon the standard pressure given above, and are for the same latitude. His Table, therefore, will give the temperature on the thermometric scale corresponding to the pressure.

The Commissioners appointed by the British Government to construct standard weights and measures, decided that the normal boiling-point, 212°, on the thermometer should represent the temperature of steam generated under an atmospheric pressure equal in inches of mercury, at the temperature of freezing water, to 29·922 + (cos. 2 latitude × ·0766) + (·00000179 × height in feet above the sea-level). Hence, at London, lat. 51°30′ N., we deduce 29·905 as the barometric pressure representing the normal boiling point of water,—the trifling correction due to height being neglected. If then, in the latitude of London, the barometric pressure, at the time of fixing the boiling point, be not 29·905 inches, that point will be higher or lower, according to the difference of the pressure from the normal. Near the sea-level about 0·59 inch of such difference is equivalent to 1° Fahrenheit in the boiling point.

Suppose, then, the atmospheric pressure at London to be 30·785 inches, the following calculation gives the corresponding boiling temperature for Fahrenheit's scale :—

Observed pressure	...	...	30·785
Normal ,,	...	...	29·905
Difference	...	...	...	·880

As 0·59 is to 0·88, so is 1° to 1°·5.

That is, the water boils at 1°·5 above its normal temperature; so that, in this case, the normal temperature to be placed on the scale, viz. 212°, must be 1°·5 lower than the mark made on the tube at the height at which the mercury stood under the influence of the boiling water.

The temperature of the vapour of boiling water may be found, at any time and place, as follows :—Multiply the atmospheric pressure by the factor due to the latitude, given in the annexed Table V., and with the result seek the temperature in Table VI.

TABLE V.   TABLE VI.

Latitude.	Factor.	Temperature of Vapour.	Tension.	Temperature of Vapour.	Tension.
Degrees.		Degrees.	Inches.	Degrees.	Inches.
0	0·99735	179	14·934	197	22·036
5	0·99739	180	15·271	198	22·501
10	0·99751	181	15·614	199	22·974
15	0·99770	182	15·963	200	23·456
20	0·99797	183	16·318	201	23·946
25	0·99830	184	16·680	202	24·445
30	0·99868	185	17·049	203	24·952
35	0·99910	186	17·425	204	25·468
40	0·99954	187	17·808	205	25·993
45	1·00000	188	18·197	206	26·527
50	1·00046	189	18·594	207	27·070
55	1·00090	190	18·998	208	27·623
60	1·00132	191	19·409	209	28·185
65	1·00170	192	19·828	210	28·756
70	1·00203	193	20·254	211	29·335
75	1·00230	194	20·688	212	29·922
80	1·00249	195	21·129	213	30·515
		196	21·578	214	31·115

How to use the Tables.—When the *temperature* is known to decimals of a degree, take out the tension for the degree, and multiply the difference between it and the next tension by the decimals of the temperature, and add the product to the tension, for the degree.

Required the tension corresponding to 197°·84.

$$197 \quad \ldots \quad \ldots = 22·036 \qquad ·465 \times ·84 = ·391$$
$$198 \quad \ldots \quad \ldots = 22·501 \qquad 197° \qquad = 22·036$$

$$\text{Difference} \quad \ldots \qquad ·465 \qquad 197·84 \qquad = 22·427$$

When the *tension* is given, take the difference between it and the next less tension in the Table, and divide this difference by the difference between the next less and next greater tensions. The quotient will be the decimals to add to the degree opposite the next less tension.

Thus, for 23·214 inches, required the temperature.

Given	23·214	Next greater	... ...	23·456
	22·974	Next less	... ...	22·974
	·240	Difference	...	·482

$$\text{And } \frac{·240}{·482} \quad ...\quad ... \quad = \quad ·5$$

Temperature opposite next less	... 199·0
Temperature required	... 199·5

A similar method of interpolation in taking out numerical quantities is applicable to almost all tables; and should be practised with all those given in this work.

Example.—Thus, in Liverpool, lat. 53° 30′ N., the barometer reading 29·876 inches, its attached thermometer 55°, and the correction of the instrument being + ·015 (including index error, capillarity and capacity), what temperature should be assigned for the boiling point marked on the thermometer?

Observed barometer	...	...	...	29·876
Correction	...	...	...	+ ·015
				29·891
Correction for temperature ...		...	...	— ·074
Reduced reading	...	...	...	29·817
Factor from Table V.	...	...	...	1·00077
				208719
				208719
			29817	
Equivalent for lat. 45°	...		...	29·83995909

In Table VI., 29·84 gives temperature 211°·86.

58. Displacement of the Freezing Point.—Either the prolonged effect of the atmospheric pressure upon the thin glass of the bulbs of thermometers, or the gradual restoration of the equilibrium of the particles of the glass after having been greatly disturbed by the operation of boiling the mercury, seems to be the cause of the freezing points of standard thermometers reading from a few tenths to a degree higher in the course of some years, as has been repeatedly observed. To obviate this small error, it is our practice to place the tubes aside for about six months before fixing the freezing point, in order to give time for the glass to regain its former state of aggregation. The making of accurate thermometers is a task attended with many difficulties, the principal one being the liability of the zero or freezing point varying constantly, so much so, that a thermometer that is perfectly correct to-day, if immersed in boiling water, will be no longer accurate; at least, it will take some time before it again settles into its normal state. Then, again, if a thermometer is recently blown, filled, and graduated immediately, or, at least, before some months have elapsed, though every care may have been taken with the production of the instrument, it will require some correction; so that the instrument, however carefully made, should from time to time be plunged into finely-pounded ice, in order to verify the freezing point.

59. The Scale.—The two fixed points having been determined, it is necessary to apply the scale. The thermometers in general use in the United Kingdom, the British Colonies, and North America are constructed with Fahrenheit's scale. Fahrenheit was a philosophical instrument maker of Amsterdam, who, about the year 1724, invented the scale which has given his name to the thermometer. The freezing point is marked 32°, the boiling point 212°, so that the intermediate space is divided into 180 equal parts, called degrees. "The principle which dictated this *peculiar division* of the scale is as follows:—When the instrument stood at the greatest cold of Iceland, or 0 degree, it was computed to contain 11124 equal parts of quicksilver, which, when plunged in melting snow, expanded to 11156 parts; hence the intermediate space was divided into 32 equal portions, and 32 was taken as the freezing point of water: when the thermometer was plunged in boiling water, the quicksilver was expanded to 11336; and therefore 212° was marked as the boiling point of that fluid. In *practice*, Fahrenheit determined the divisions of his scale from two fixed points, the freezing and boiling of water. *The theory* of the division, if we may so speak, was derived from the lowest cold observed in Iceland, and the expansions of a given portion of mercury" *(Professor Trail)*.

The divisions of the scale can be carried beyond the fixed points, if requisite, by equal graduations. Fahrenheit's scale is very convenient in some respects. The meteorological observer is seldom troubled with negative signs, as the zero of the scale is much below freezing. Again, the divisions are more numerous, and consequently smaller, than on other scales in use; and the further subdivision into tenths of degrees, seems to give all the minuteness usually required.

Celcius, a Swede, in 1742, proposed zero for the freezing point, and 100 for the boiling point, all temperatures below zero being distinguishable by the sign (—) minus. This scale is known as the *centigrade*, and is in use in France, Sweden, and the southern part of Europe. It has the advantage of the decimal notation, with the embarrassment of the negative sign.

Reaumur, a Frenchman, proposed zero for the freezing point, and 80° for the boiling point, an arrangement inferior to the centigrade. It is, however, in use in Spain, Switzerland, and Germany.

It is merely a simple arithmetical operation to change the indications of any one of these scales into the equivalents on the others. To facilitate such conversions, tables are convenient, when a large number of observations are under discussion; and they can be easily formed or obtained.

In the absence of such tables, the following formulæ will insure accuracy of method, and save thinking, when occasional conversions are wanted to be made:— F. stands for Fahrenheit, C. for Centigrade, and R. for Reaumur.

Given.			Required.			Solution.
F.	...	...	C.	...	...	$= \quad (F. - 32) \frac{5}{9}$
F.	...	...	R.	...	...	$= \quad (F. - 32) \frac{4}{9}$
C.	...	...	F.	...	...	$= \quad \frac{9}{5} C. + 32$
C.	...	...	R.	...	...	$= \quad \frac{4}{5} C.$
R.	...	...	F.	...	...	$= \quad \frac{9}{4} R. + 32$
R.	...	...	C.	...	...	$\quad \frac{5}{4} R.$

Example.—Convert 25° of Fahrenheit's scale into the corresponding temperature on the Centigrade scale.

$$\text{Here } C. = (25 - 32) \frac{5}{9}$$
$$C. = - \frac{3\cdot5}{9} = - 3\cdot9$$

or nearly 4° *below* zero of the Centigrade scale. The algebraical sign must be carefully attended-to in the calculations.

60. The method of testing Thermometers or meteorological purposes is very simple. Such thermometers are seldom required to read above 120°. In these the freezing point having been determined, the divisions of the scale are ascertained by careful comparisons, with a standard thermometer, in water of the requisite temperature. "For the freezing point, the bulbs, and a considerable portion of the tubes of the thermometers, are immersed in pounded ice. For the higher temperatures, the thermometers are placed in a cylindrical glass vessel containing water of the required heat: the scales of the thermometers intended to be tested, together with the Standard with which they are to be compared, are read through the glass. In this way the scale readings may be tested at any required degree of temperature, and the usual practice is to test them at every ten degrees from 32° to 92° of Fahrenheit."—*FitzRoy*.

61. Porcelain Scale Plates.—Thermometer scales of brass, wood, or ivory, either by atmospheric influence or dipping in sea-water, are very liable to become soiled and discoloured, so much so that after a very little time the divisions are rendered nearly invisible. To obviate this inconvenience, Messrs. Negretti and Zambra were the first to introduce into extensive use thermometer and barometer scale-plates made of porcelain, having the divisions and figures engraved thereon by means of fluoric acid, and permanently burnt-in and blackened, so as always to present a clear legible scale. That these scales have been found superior to all others, may be inferred from the fact that all the thermometers now supplied to the various government departments are provided with such scales.

They can be adapted to replace any of the old forms of brass or zinc scales, the divisions and figures of which have become obliterated or indistinct.

62. Enamelled Tubes.—Nearly all thermometer tubes are now made with enamelled backs. This contrivance of enamelling the backs of the tubes enables the makers to use finer threads of mercury than had before been found practicable ; for were it not for the great contrast between the dark thread of mercury and the white enamel on the glass, many of the thermometers now in use would be positively illegible. The enamelling of thermometers is an invention of Messrs. Negretti and Zambra. It is necessary to state this, as many persons, from interested motives, are anxious to ignore to whom the credit of the invention is due.

63. Thermometers of extreme Sensitiveness.—Thermometers for delicate experiments are no novelty. Thermometers have been made with very delicate bulbs to contain a very small quantity of mercury. Such instruments have also been made with spiral or coiled tubular bulbs, but the thickness of glass required to keep these coils or spirals in shape, and in fact to prevent their falling to pieces, served to nullify the effect sought to be produced, viz. instantaneous action ; and where a small thin bulb was employed, the indicating column was generally so fine that it was positively invisible except by the aid of a powerful lens. Messrs. Negretti and Zambra have now introduced a new form of thermometer, which combines sensitiveness and quickness of action, together with a good visible column. The bulb of this thermometer is of the gridiron form. Care has been taken in constructing the bulb, so that the objections attending spirals and other forms have been overcome ; for whilst the reservoir or bulb is made of glass so thin that it is only by a spirit lamp and not a glass blower's blowpipe that it can be formed, yet it is still so rigid (owing to its peculiar configuration) that no variations in its indications can be detected, whether it be held in a horizontal, vertical, or oblique position, nor will any error be detected if it be stood on its own bulb. They have made thermometers with bulbs or reservoirs formed of about nine inches of excessively thin cylindrical glass, whose outer diameter is not more than a twentieth of an inch; so that, owing to the large surface presented, the indications are positively instantaneous. This form of thermometer was constructed expressly to meet the requirements of scientific balloon ascents, to enable thermometrical readings to be taken at the precise elevation. It was contemplated to procure a metallic thermometer, but on the production of this perfect instrument the idea was abandoned.

64. VARIETIES OF THERMOMETERS.

Fig. 37 is an illustration of boxwood scale thermometers for general use and common purposes.

Fig. 38, Negretti and Zambra's Travelling Thermometer; it is fixed in a plated metal (silver or otherwise) case, similar to a pencil-case, and has the scale divided upon its stem.

Fig. 39, Thermometer mounted on a slab of glass, upon which the scale is etched, the back being either oak, mahogany, or ebony.

Fig. 40, Portable Thermometer, in a bronzed brass or German silver revolving case.

Fig. 41, Pocket Thermometer, on ivory or metallic scale, in morocco or papier-mâché case.

Fig. 37.   **Fig. 38.**   **Fig. 39.**   **Fig. 40.**   **Fig. 41.**

    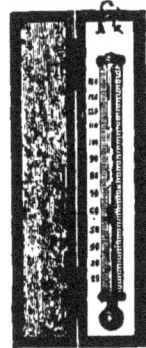

Fig. 42, an Ornamental Drawing-room Thermometer, on ebony or ivory stand, with glass shade.

Fig. 43, representation of highly carved or engine-turned design for thermometer mounts, in ivory or wood, for the drawing-room. Some have the addition of a sun-dial or compass at the top; they may also be formed for a watch-stand.

Fig. 44, **Bath Thermometer,** having a float to admit of its being kept in the water.

Fig. 42.   **Fig. 43.**   **Fig. 44.**

 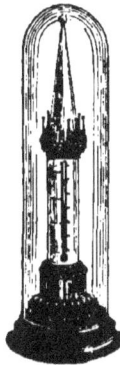 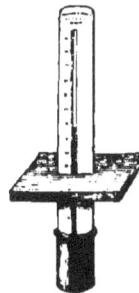

Fig. 45, Thermometer with ivory scale in glass cylinder, mounted on oak bracket with metal top, for out-door use; as at a window.

Fig. 46, Thermometer for the window, on patent porcelain or glass scale, with oak bracket and convenient brass supports, for placing the instrument at any angle.

Fig. 47, **Chemical Thermometer,** on boxwood scale, jointed near the bulb on a brass hinge, ranging from 300° to 600°.

Fig. 48, **Chemical Thermometer,** for acids, graduated on its own stem, suitable for insertion in the tubulure of retorts; they are also made insulated in glass cylinder to protect the graduated stem; ranging from 0° to 600°.

Fig. 45.   **Fig. 46.**   **Fig. 47.**   **Fig. 48.**   **Fig. 49.**   **Fig. 50.**

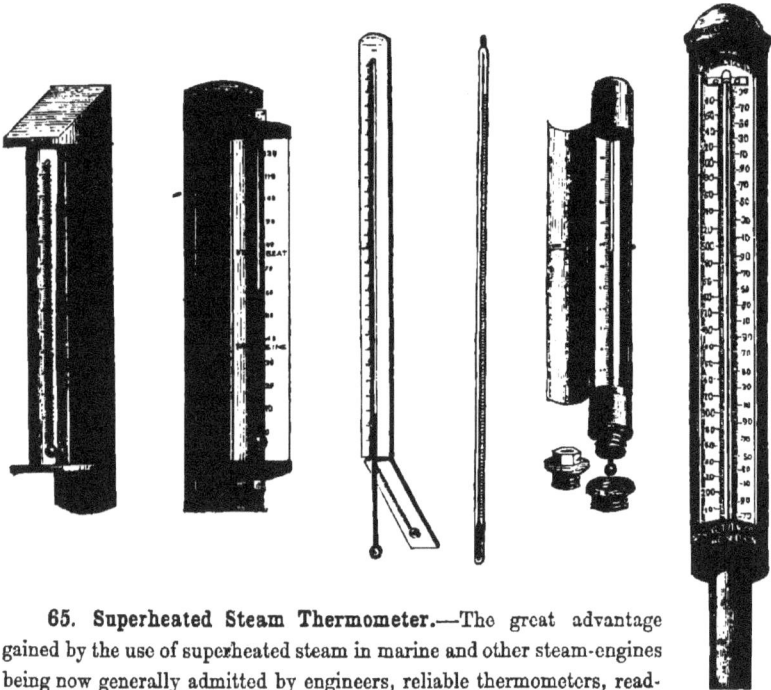

65. Superheated Steam Thermometer.—The great advantage gained by the use of superheated steam in marine and other steam-engines being now generally admitted by engineers, reliable thermometers, reading to 600° at least, are of the utmost importance. To meet this want, Messrs. Negretti and Zambra have constructed for the purpose a substantial form of thermometer, on their patent porcelain scales, in strong and convenient metal mountings, with perforated protection to the bulb. The scales cannot be deteriorated by steam, heat, oil, or dirt; and an occasional wiping will be all that is necessary to keep the divisions and figures clean and visible for any length of time; while careful calibration of the thermometer tubes ensures the most accurate indications attainable. These thermometers are illustrated by figs. 49 & 50. A similar, but cheaper, construction is given to thermometers to be used with hot air, or hot water, apparatus.

66. Thermometer for Sugar Boiling is protected by a metallic frame; and is usually from three to four feet long, the graduations being confined to a space of about twelve inches at the upper part of the instrument, allowing the bulb and greater part of the tube to be immersed in the boiling sugar. The graduations extend to 270° or further. An index is sometimes attached to the scale, which may be set to any degree of heat required to be maintained.

67. EARTH THERMOMETER.

The Earth Thermometer is for ascertaining the temperature of the soil at various depths. It is protected by a brass frame, pointed and strengthened at the end to facilitate insertion into the ground, as in fig. 51.

Utility of a Knowledge of the Temperature of the Soil.—The temperature of the soil is an important element in the consideration of climate, as it concerns the vegetable kingdom.

Fig. 51.

Dr. Daubeny, in his *Lectures on Climate*, gives the following statement with respect to some temperatures which have been observed just beneath the earth's surface, in different parts of the globe:—

Country.	Temperature.	Authority.
Tropics, often	162-134°	Humboldt.
Egypt	133-144...........................	Edwards & Colin.
Orinoco	In white sand, 140..............	Humboldt.
Chili	113-118, among dry grass......	Boussingault.
Cape of Good Hope...	150, under the soil of a bulb garden	Herschell.
Bermuda	142, thermometer barely covered in earth	Emmet.
China	Water of the fields, 113; adjacent sand, much higher; blackened sides of the boat at midday, 142-150	Meyer.
France	118-122, and in one instance 127	Arago.

"The importance of this to vegetation may be estimated by the following considerations:—

"It is known that every plant requires a certain amount of heat, varying in the case of each species, for the renewal of its growth, at the commencement of the season.

"Now when this degree of heat has spurred into activity those parts that are above ground, and caused them to elaborate the sap, it is necessary that the subterranean portions should at the same time be excited by the heat of the ground to absorb the materials which are to supply the plant with nourishment. Unless the latter function is provided for, the aerial portions of the plant will languish from want of food to assimilate. Indeed, it is even advisable that the roots should take the start of the leaves, in order to have in readiness a store of food for the latter to draw upon." In another place the professor remarks:—"It has been calculated by Mr. Raikes, from experiments made at Chat Moss, that the temperature of the soil when drained averages 10° more than it does when undrained; and this is not surprising, when we find that 1 lb. of water evaporated from 1,000 lbs. of soil will depress the whole by 10°, owing to the latent heat which it absorbs in its conversion into vapour."

68. MARINE THERMOMETER.

Fig. 52.

This instrument is a special construction to meet the requirements of navigation. It consists of a carefully constructed thermometer divided on its stem to degrees, which are sufficiently large to admit of subdivision into tenths of degrees by estimation, and ranging from 0° to 130°. The scale is porcelain, having the degrees etched upon it, and burnt-in a permanent black. The instrument is made to slide into a japanned metallic case, for handy use and protection. It is therefore adapted for almost any ordinary purpose; and cannot be injuriously affected by any chemical action arising from air or sea-water. A set of these thermometers consists of six, carefully packed in a neat box; two having japanned metallic cases (fig. 52), the others being designed for use without the case, or to replace a breakage.

This thermometer is employed in the Royal Navy, and for the observations made at sea for the Board of Trade.

The thermometer is now considered a necessary instrument on board ship. Not only is it of invaluable utility in connection with the barometer as a guide to the weather, but its indications are of service in showing the presence of a warm or cold current in the sea; many of the great oceanic currents being characterised by the warmth or coldness of their waters. In seas visited by icebergs, the habitual use of the thermometer would indicate their proximity, as the water is rendered colder for some distance around by the thawing of huge masses of ice. The water over a shoal in the sea is generally colder than the surface-water of the surrounding ocean; which may result from the cold water being brought to the surface by the current of water encountering the shoal. With this fact navigators are well acquainted; and therefore a fall in the sea-water thermometer may forebode that shallow water is at hand. It has been ascertained that fish inhabit regions of the oceans and seas having the peculiar temperature suitable to their habits. The better and firmer sort of fish are found where cold waters exist. Those taken in warmer belts or streams of water, even in the same latitude, are far inferior in condition, and less approved by the palate. The fish of the Mediterranean, a warm sea, are generally poor and scarce. Fish taken in the cold waters between the American shore and the Gulf Stream are much esteemed; while in and on the other side of the stream they are said to be tasteless, and of no flavour. Between the coasts of China and the warm waters of the Japanese current, the seas abound with excellent fish; but in the warm waters of the current and beyond, they are never seen in such shoals.

. In fact, it is clearly ascertained that fishes are adapted to climates, like birds and beasts. It has been even affirmed, after careful investigation, that herrings, which abound in the British Seas, and form a most important branch of our fisheries, can only be found in a temperature varying from 54° to 58°. Hence the thermometer, if brought into use by the fishermen, would guide them to the spots where they may with the best chance cast their nets on dark nights, when other indications are not perceptible.

This thermometer in its metallic case is perfectly suited for dipping overboard, or placing in a bucket of water just taken from the sea, to ascertain its temperature.

CHAPTER VII.

69. Importance of Self-Registering Thermometers.—Heat being apparently the most effective agent in producing meteorological phenomena, the determination of the highest temperature of the day, and the lowest during the night, is a prime essential to enable an estimate of the climate of any place to be formed. To observe these extremes by means of the ordinary thermometer would be impracticable, from the constant watchfulness which would be necessary. Hence, the utility and importance of self-recording thermometers are evident. A thermometer constructed to *register* the highest temperature is usually called a *maximum thermometer;* one to show the lowest temperature is termed a *minimum thermometer;* and if made to record both extremes of temperature, it is designated a *maximum-and-minimum thermometer.* We will, for the sake of method, describe the instruments in use in this order.

It would carry us beyond our scope to explain in detail the methods of dealing with temperature observations; but we may remark that half the sum of the maximum and minimum temperature of each day of twenty-four hours, is not what meteorologists designate the *mean daily temperature*, although it very frequently approximates to it. The mean temperature of the day is understood to be the average of twenty-four consecutive hourly readings of a thermometer; and meteorology now supplies formulæ whereby this result can be deduced from two or three observations only in a day. But we would observe that the actual mean temperature of any place has not such an important influence upon life, either animal or vegetable, as the abruptness and magnitude of the variations of temperature. Climate, therefore, should be estimated more by the range of the thermometer than by the average of its indications. The Registrar General's returns prove that with a wide range of the thermometer, the mortality greatly increases; and it is now becoming apparent to meteorologists that the daily range of the thermometer marks the effects of temperature on the health of men, and the success of crops, better than any other meteorological fact of which we take cognizance. Now that self-registering thermometers are constructed with mercury, the most appropriate of all thermometric substances, not only for maxima, but likewise for minima temperatures, the determination of the diurnal range of temperature is rendered more certain, and observations at different places are more strictly comparable.

MAXIMA THERMOMETERS.

70. Rutherford's Maximum Thermometer.—The maximum thermometer, invented by Dr. John Rutherford, differs from an ordinary thermometer in having a small cylinder of steel, porcelain, or aluminium, moving freely in the tube beyond the mercury, so as to form an index. The stem of the thermometer is fixed horizontally on the frame, which must be suspended in the same position, as represented in fig. 53.

Fig. 53.

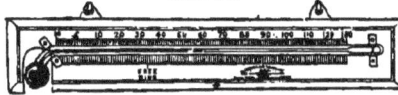

The instrument is set by holding it bulb downward, so as to allow the index to fall by its own gravity into contact with the mercury. Increase of heat produces expansion of the mercury, which consequently pushes forward the index. When the temperature decreases, the mercury recedes from the index, leaving it so that the extremity which was in contact with the mercury indicates upon the scale the highest temperature since the instrument was last set.

As it is easily constructed and is comparatively cheap, it is still employed for ordinary purposes. Its disadvantages are, firstly, its liability of soon getting out of order by the index becoming embedded in the mercury, or fixed by oxidation, thus rendering it altogether useless; secondly, the ease with which the index can be displaced by the wind moving the instrument, or other accidental disturbance, so as to cause it to give erroneous indications occasionally; and thirdly, its consequent total unfitness for use at sea.

In the part of the tube beyond the mercury, a small quantity of air is enclosed for the purpose of preventing the metal flowing freely in the tube. This necessitates the construction of a larger bulb, which renders the thermometer less sensitive. Moreover, as it frequently happens that some mercury passes the index, particles of air insinuate themselves in the metal, and cause separations in the column, which very often can be removed only by a maker. To facilitate this re-adjustment, a small chamber is left at the end of the tube, and the mercury being expanded into it by heat until the index and air bubbles are forced into it, if possible, upon the cooling down again, by a little management, the mercury will contract, leaving the air and index behind. Yet sometimes the index cannot be moved in the least from its place of fixture, so that the instrument must be virtually reconstructed.

71. Phillip's Maximum Thermometer.—A maximum thermometer, better perhaps in its action than Rutherford's, has been suggested by Professor John Phillips, of Oxford. A small portion of air is introduced into an ordinary thermometer, so as to cut off about half an inch of the mercurial thread near its end in the tube. This forms a maximum thermometer, when the stem is arranged horizontally. The isolated portion is pushed forward by expansion, and is left in this position when the

mercury contracts.   The end remote from the bulb shows on the scale the maximum temperature.

When made with a capillary tube so fine that the attraction arising from
capillarity overcomes the force of gravity, and prevents the mercury falling to the
end of the tube when the instrument is inverted, it forms a very serviceable thermometer, quite portable and suitable for use on board ship.   In such a tube a
smart shake from a swing of the hand is required to bring the detached portion
back to the column, so as to set the instrument for future observation ; no ordinary
motion will move it.   When the thermometer has not this peculiarity, the mercury
will flow to the end, if held bulb downward ; and in this state it is not at all a satisfactory instrument, as the air is likely to be displaced, and a great deal of tact is
requisite to again get it to divide the column suitably.   It has been found in practice
that the air bubble at different temperatures assumes different lengths, and if very small
it disappears in a few years by oxidation and by diffusion with the mercury, so that
the instrument becomes defective and uncertain in action,—results which led to the
construction of the self-registering mercurial maximum thermometer, invented and
patented by Messrs. Negretti and Zambra.   It has been before the public about
twelve years; we may therefore, now, safely speak of its merits.

72. **Negretti and Zambra's Patent Maximum Thermometer** consists of a glass
tube containing mercury fitted on an engraved scale, as shown in fig. 54.   The part

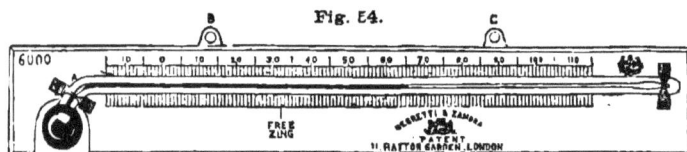

Fig. 54.

of the thermometer tube above the mercury is entirely free from air; and at the
point ▲ in the bend above the bulb, is inserted and fixed with the blow-pipe a
small piece of solid glass, or enamel, which acts as a valve, allowing mercury to
pass on one side of it when heat is applied, but not allowing it to return when the
thermometer cools.   When mercury has been once made to pass the contraction,
which nothing but the expansive force of heat can effect, and has risen in the tube,
the upper end of the column registers the maximum temperature.   To return the
mercury to the bulb, we must apply a force equal to that which raised it in the
tube; the force employed is gravity, assisted when necessary by a little agitation of
the instrument.

The degrees are generally divided on the stems of these thermometers, but
their frames of course bear a scale as well.   The makers have various styles of
framing in wood, metal, porcelain, and even glass.   Each material is eligible according to requirements.   Porcelain scales, having the marks *etched* upon them by
acid and permanently blackened and baked in,—by a process for which the

inventors have a separate patent,—will be found very serviceable, as they do not corrode or tarnish by exposure to any kind of weather; while any amount of dust and dirt can readily be cleaned off.

The chief recommendation of this thermometer is its simplicity of construction, enabling it to be used with confidence and safety. Of no other maximum thermometer can it be said that it is impossible to derange or put it out of order; hence, as regards durability, it surpasses all others. Nothing short of actual breakage can cause it to fail. Hence it is the most easily portable of all self-registering thermometers, an advantage which renders it suitable for travellers, and for transmission abroad. In the year 1852, the British Meteorological Society reported this thermometer to be "the best which has yet been constructed for maximum temperature, and particularly for sun observations." Since then eleven years have elapsed, and it is still without a rival.

Directions for use: In using this thermometer for meteorological observations, it should be suspended by means of two brass plates B, C, attached for that purpose, in such manner that it hangs raised up a little at C, and so placed that it is in the shade, with the air passing freely to it from all sides; then, on an increase of heat, the mercury will pass up the tube as in an ordinary thermometer, and continue doing so as long as the heat increases. On a decrease of heat, the contraction of mercury will take place *below* the *bend* in the tube, leaving the whole column of mercury in the tube, thus registering the highest temperature, and showing such till the instrument is disturbed.

To prepare the instrument for future observations, remove and hold it perpendicularly, with the bulb downward, and then shake it. The mercury will then descend in the tube, and indicate the temperature of the air at that time; and, when again suspended, is prepared for future observation.

After the temperature has attained a maximum, there will be, with a decrease of heat, a slight contraction of mercury in the tube—as well as of that in the bulb —and hence doubts have arisen as to the accuracy of the registration; but calculation shows, and critical trial has proved, that the greatest daily range of temperature will not produce an error large enough to be appreciable on the scale.

A very great advantage of this thermometer is that the mercury may be allowed to flow to the end of the tube without the maximum temperature attained during an experiment being lost. It can be employed with the bulb uppermost. All that is necessary for reading the maximum temperature is to slope the instrument so that the mercury flows gently towards the bulb. It will then stop at the contraction so as to show the maximum temperature on the scale. Afterwards the mercury is driven into the bulb by agitating the instrument while held in the hand. Hence the instrument is invaluable as a registering thermometer on board ship, as its indications are in no way affected by the motions and tremors of the vessel.

For physiological experiments, such as taking the temperature of the mouth in fever, this thermometer is the only one that can be used with certainty, as it can be held in any position, without losing the maximum temperature attained.

MINIMA THERMOMETERS.

73. Rutherford's Alcohol Minimum Thermometer, fig. 55, consists of a glass tube, the bulb and part of the bore of which is filled with perfectly pure spirits of wine, in which moves freely a black glass index. A slight elevation of the thermometer, bulb uppermost, will cause the glass index to flow to the surface of the

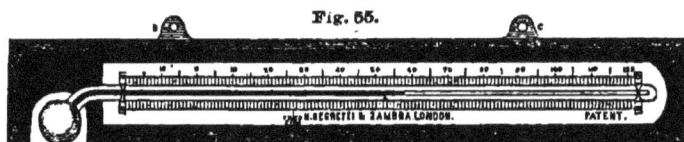

Fig. 55.

liquid, where it will remain, unless violently shaken. On a *decrease* of temperature the alcohol recedes, taking with it the glass index; on an *increase* of temperature the alcohol alone ascends in the tube, leaving the end of the index *farthest* from the bulb indicating the minimum temperature.

Directions for using, &c.—Having caused the glass index to flow to the end of the column of spirit, by slightly tilting the thermometer, bulb uppermost, suspend the instrument (in the shade with the air passing freely to it on all sides) by the two brass plates attached for that purpose,—in such manner that the bulb is about half an inch lower than the upper, or the end of the thermometer farthest from the bulb; then, on a decrease of temperature, the spirits of wine will descend, carrying with it the glass index; on an increase of temperature, however, the spirits of wine will ascend in the tube, leaving that end of the small glass index farthest from the bulb indicating the minimum temperature. To reset the instrument, simply raise the bulb end of the thermometer a little, as before observed, and the index will again descend to the end of the column, ready for future observation.

Precautions.—1. By no means jerk or shake an alcohol minimum thermometer *when resetting* it, for by so doing it is liable to disarrange the instrument, either by causing the index to leave the spirit, or by separating a portion of the spirit from the main column.

2. As alcohol thermometers have a tendency to read lower by age, owing to the volatile nature of the fluid allowing particles in the form of vapour to rise and lodge in the tube, it becomes necessary to compare them occasionally with a mercurial thermometer whose index error is known; and if the difference be more than a few tenths of a degree, examine well the upper part of the tube to see if any alcohol is hanging in the bore thereof; if so, the detached portion of it can be joined to the main column by swinging the thermometer with a pendulous motion, *bulb downwards.*

3. The spirit column is sometimes much separated by jolting in travelling. If the instrument is in such a condition when received, it should be held by the right

hand, bulb downward, and the frame tapped smartly, but cautiously, against the palm of the left hand. The broken thread of spirit will soon begin to join, and by continuing the operation a sufficient time all the bubbles will disappear, and the thermometer become as good as ever.

74. Horticultural Minimum Thermometer.—This instrument, represented in fig. 56, is a special construction of Rutherford's minimum thermometer to meet the requirements of horticulturists. It is desirable, if not essential, that gardeners

Fig. 56.

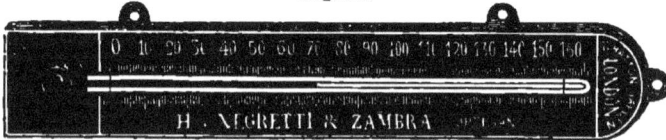

should have the means of ascertaining to what temperature stoves and greenhouses descend on cold nights, especially in winter. This thermometer is mounted on a strong cast zinc frame, with the divisions and figures of the scale raised.

The sunk surface of the frame is painted dark; the figures and division a bright colour, so that observations can be made without a close inspection of the instrument.

The directions for using are the same as those given in the preceding section. It may be used as an ordinary thermometer, by simply hanging it from the top loop, in which position, the coloured liquid will always indicate the present temperature.

It was a source of annoyance with the ordinary boxwood and flat metal scales, that after a time, exposure to a damp warm atmosphere favoured the growth of confervæ upon them, and obliterated the divisions; the plan of raising the figures and divisions of the scale has been found to prevent the destruction of the instrument in this way.

75. Baudin's Alcohol Minimum Thermometer.—This instrument resembles Rutherford's thermometer in appearance; its indications are given by the expansion and contraction of alcohol, and its minimum temperature is likewise registered by a glass index being pulled back and left behind by the alcohol, as in Rutherford's instrument. There is, however, a great improvement in Baudin's instrument; for whilst Rutherford's thermometer can only register in a horizontal position, Baudin's can be used either horizontally or vertically, as necessity may require. This important change is effected in the following manner:—Instead of the index in the thermometer being loose and free to run up and down according to the position in which the instrument is held, as in Rutherford's, the index in the new instrument is made to fit the bore of the tube as nearly tight as possible, so much so that in holding the thermometer even upside down, or shaking it, the index will not shift from its position; but, inasmuch as a minimum thermometer with an immoveable index could not be set when required for observation, and would consequently be useless, the inventor has introduced behind the index a piece of solid glass, about one-and-a-half inch in length, which moves freely in the alcohol. The addition of the weight of this piece of glass on the top of the index, when turned upside down, forces the index down to the edge of the alcohol; and it is there left, as in the case of the ordinary Rutherford's thermometer. It is, therefore, by turning the thermometer upside down, and letting the moveable piece of glass fall on the index, that the index is driven to the end of the alcohol; after this operation the ther- mometer is hung up either horizontally or vertically, and will then be ready for use.

The index, although immoveable *per se*, is by the alcohol drawn back, as in the ordinary minimum, and its indications are read off on the scale from the top of the index.

Fig. 57.

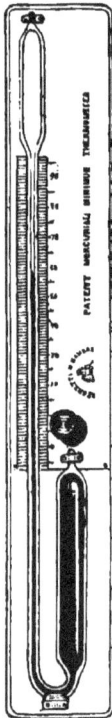

76. Mercurial Minima Thermometers desirable.—Alcohol does not expand equally for equal increments of heat, consequently errors are likely to exist in the scale indications unless the graduations are very accurately—not necessarily equally—made. On this account, as well as from the volatility of alcohol, and the intervention of gaseous partitions in the tube, a good and thoroughly reliable minimum ther- mometer was for a long time a desideratum. It was desirable to obtain a thermometer which should register the lowest temperature by mercury, the fluid in general use for meteorological thermometers. Several instruments have recently been invented to meet this require- ment, which are suitable and satisfactory for land purposes, but one well adapted for use on board ship is still very much wanted.

For very low temperatures, alcohol thermometers will always be required; as mercury freezes at − 40° F, and contracts very irregu- larly much before this point, while alcohol has never yet been frozen.

77. Negretti and Zambra's Patent Mercurial Minimum Thermometer, represented by fig. 57, has a cylindrical bulb of large size, which, at first sight, might induce the idea that the instrument would not be sufficiently sensitive; but as length is given to the cylinder instead of increasing its diameter, it will be found as sensitive as a globular bulb of the same diameter, and much more so than an ordinary alcohol thermometer.

The reason for having the bulb large is to allow the internal diameter of the thermometer tube to be larger than that generally used for thermometrical purposes, so that a steel index, pointed at both ends, may move freely within when required.

The tube is blown, filled and regulated in the usual way, 60° of temperature being about half-way up the tube. A small cylindrical bulb is then formed at the upper end of the tube, and then is introduced a steel needle pointed at both ends, that in contact with the mercury being abrupt, the other more prolonged. The open extremity of the tube is now drawn out into a fine capillary tube, and the bulb of the instrument warmed so as to cause the mercury to fill the tube completely. When the mercury reaches the capillary tube, the flame of a blow-pipe is applied ; the glass is dexterously melted, the superfluous part taken away, and the tube left hermetically closed. During this operation, the steel index has been embedded in the heated mercury. As the instrument cools, if held upright, the mercury will recede and expose the needle, which will then follow the descending column simply by its own gravity. In this condition the thermometer resembles Rutherford's maximum, being a tube of mercury with a steel index floating on its surface; but it possesses these important advantages : it is quite free from air, so that the mercury can move with perfect freedom; and the index is pointed at both ends, to allow the mercury to pass, instead of being ground flat to prevent it.

To use the Thermometer, it is suspended perpendicularly (figure 57) with the steel index resting on the surface of the mercurial column. As the mercury in the cylinder contracts, that in the tube descends, and the index, of its own gravity, follows it; on the contrary, as the mercury expands and rises in the tube, it passes the index on one side, and in rising, exerts a lateral pressure on the needle, and jams it to one side of the tube, where it remains firmly fixed, leaving the upper point of the needle indicating the minimum temperature. In this thermometer, the reading is always from the upper point of the needle, and not from the mercury itself.

To extricate the Needle from the mercury, a magnet is used, when, if the needle is embedded only a few degrees, it can readily be withdrawn without altering the position of the instrument. Should the magnet not be sufficient for the purpose, we simply turn the thermometer on its support from the upright position, slightly elevating the bulb (fig. 58 (2)). The mercury and index will then flow into the small reservoir. Should the index not freely leave the tube with the mercury, assist it with a magnet, and when the mercury and index are in the upper bulb (figure 2), apply a magnet outside, which will attract and hold fast the index; and whilst thus holding it, again bring the thermometer to the upright position, when the mercury will immediately fall back into the tube, leaving the index attached to the magnet (figure 4), with which it is guided down to the surface of the mercury, ready for another observation.

Care must be taken not to withdraw the magnet until the index is in contact with mercury; for, if released before touching, it might plunge too deeply, and give a false indication. The rule for re-setting it will be to bring the needle-point in contact with the mercury, and then withdraw the magnet, having previously ascertained that no particles of mercury are attached to the index.

It may sometimes, though rarely, happen, that from the time a

Fig. 58.

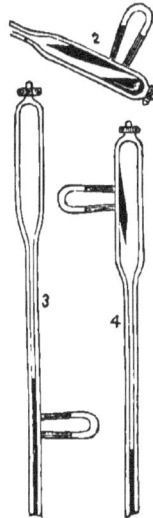

minimum temperature is registered by the index, and by the time an observation is made, the mercury may have risen so high in the tube as to completely pass the index, as shown (figure 3). Should it so happen, the space which the index occupies will readily be observed, as it will be pressed to one side of the tube, causing a different appearance in that part, although the point of the needle may not be seen. If such be the case, apply a magnet to the spot where you see the index is fixed: this will hold the needle firmly. Then, by slightly tilting the thermometer bulb uppermost, the mercury will flow into the top bulb, leaving the index attached to the magnet, and quite uncovered. Having taken the reading, draw the needle into the top bulb, and hold it there whilst you adjust the thermometer by again bringing it to the upright position.

By contracting the bore of this thermometer, at the bend of the tube, sufficiently to keep the mercury from flowing out of its bulb with too much freedom by motion, the instrument becomes perfectly safe for transmission abroad.

78. Negretti & Zambra's Second Patent Mercurial Minimum Thermometer.— In this thermometer a principle is used that has been long known to scientific men, viz. the affinity of mercury for platinum. If mercury be placed in contact with platinum under ordinary circumstances, no effect will take place; but if the mercury is once made to attack the platinum, the amalgamation is permanent and the contact perfect, so much so, that the principle was made use of in constructing standard barometers. A ring of platinum was fused round the end of the tube, dipping into the mercury; and the contact between the platinum and mercury became so perfect that air could not creep down the tube and up the bore, as in ordinary barometer tubes. This principle of adhesion or affinity of mercury for platinum has been brought into play for the purpose of arresting the mercury after it has reached the minimum temperature in a thermometer. This thermometer is made as follows:— behind the bulb is placed a supplementary chamber; in the space or neck between the bulb of the thermometer and the chamber, is placed a small piece of platinum; this may be of any shape or size, but the smaller the better. This is not to fit in the neck; it must, on the contrary, be rather loose; it may be fastened in position or not. The instrument is represented by fig. 59.

Fig. 59.

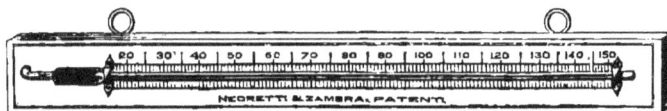

Directions for using.—Having suspended the thermometer in a horizontal position, the mercury is made to stand in exact contact with the platinum plug by slightly elevating the bulb end of the instrument. The thermometer is now ready for observation. On a decrease of temperature, the mercury will endeavour to contract first from the easier passage, viz. behind the bulb; but in consequence of the adhesion of the mercury to the platinum, it cannot recede from here, it is therefore forced to contract from the indicating tube, and will continue to do so as long as the temperature decreases; and as no indices are employed in this thermometer, the extreme end of the mercurial column will show "how cold it has been." On an increase of temperature the mercury will glide over the platinum plug and expand by the easier passage into the supplementary chamber, and there remain until a decrease of temperature again takes place, when the mercury that had gone into the supplementary chamber will be the first to recede, until it reaches the platinum plug, its further progress being arrested; it will then fall in the indicating tube, and there remain until re-set.

79. Casella's Mercurial Minimum Thermometer.—The general form and arrangement of this instrument is shown in fig. 60. A tube with large bore, *a*, has

Fig. 60.

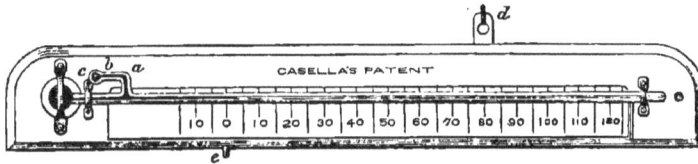

at the end a *flat glass diaphragm* formed by the abrupt junction of a small chamber, *b c*, the inlet to which at *b* is larger than the bore of the indicating tube. The result of this is that on setting the thermometer, as described below, the contracting force of the mercury in cooling withdraws the fluid in the indicating stem only; whilst on its expanding with heat, the long column does not move, the increased bulk of mercury finding an easier passage into the small pear-shaped chamber attached.

We believe that a small speck of air must be confined in the chamber, *b c*, to act as a spring to start the mercury from the chamber in the act of setting the thermometer. Were this air not present, the mercury would so adhere to the glass that no amount of shaking could induce it to flow from the chamber.

To set the Instrument, place it in a horizontal position, with the back plate, *d*, suspended on a nail, and the lower part supported on a hook, *e*. The bulb end may now be gently raised or lowered, causing the mercury to flow slowly until the bent part, *a*, is *full* and the chamber, *b c*, *quite empty*. At this point the flow of mercury in the long stem of the tube is arrested, *and indicates the exact temperature* of the bulb or air at the time. On an increase of temperature the mercury will expand into the small chamber, *b c;* and a return of cold will cause its recession from this chamber only, until it reaches the diaphragm, *b*. Any further diminution of heat withdraws the mercury down the bore to whatever degree the cold may attain, where it remains until further withdrawn by increased cold, or till re-set for future observation.

G

MAXIMA AND MINIMA THERMOMETERS.

80. Rutherford's arrangement for obtaining a complete instrument for the registration of heat and cold was simply mounting a maximum thermometer and a minimum
thermometer upon the same frame or slab.   Thus constructed, they are often called
"day and night" thermometers, though somewhat inappropriately; for in tem-
perate climates the temperature of the night sometimes exceeds that of the day,
notwithstanding the reverse is the general law of temperature.   Fig. 61 will explain
the arrangement of Rutherford's day and night thermometer.

Fig. 61.

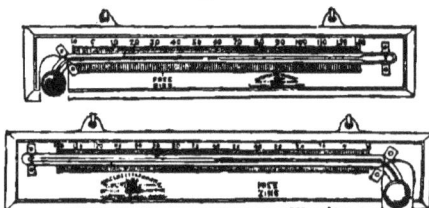

81. Size's Self-Registering Thermometer.—The very ingenious and certainly
elegant instrument about to be described was invented by James Size, of Colchester.

Fig. 62.

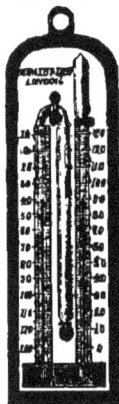

It consists of a long cylindrical bulb, united to a tube of more than
twice its length, bent round each side of it in the form of a syphon,
and terminated in a smaller, oval-shaped bulb.   Figure 62 gives a
representation of this instrument.   The lower portion of the syphon
is filled with mercury ; the long bulb, the other parts of the tube,
and part of the small bulb, with highly rectified alcohol.   A steel index
moves in the spirit in each limb of the syphon.   The two indices are
terminated at top and bottom with a bead of glass, to enable them to
move with the least possible friction, and without causing separation
of the spirit, or allowing mercury to pass easily.   They would, from
their weight, always rest upon the mercury ; but each has a fine
hair tied to its upper extremity and bent against the interior of the
tube, which acts as a spring with sufficient elasticity to keep the
index supported in the spirit in opposition to gravity.

The instrument acts as follows :—A rise of temperature causes the spirit in the
long bulb to expand and press some of the mercury into the other limb of the
syphon, into which it rises also from its own expansion, and carries the index with
it, until the greatest temperature is attained.   The lower end of this index then
indicates upon the engraved scale the maximum temperature.   As the temperature
falls the spirit and the mercury contract, and in returning towards the bulb the
second index is met and carried up by the mercury until the lowest temperature
occurs, when it is left to indicate upon the scale the minimum temperature.   The
limb of the syphon adjoining the bulb requires, therefore, a descending scale of

thermometric degrees ; the other limb, an ascending scale. The graduations must be obtained by comparisons with a standard thermometer under artificial temperatures, which should be done in this way for every 5°, in order to correct for the inequality in the bore of the tube, and the irregular expansion of the spirit. The instrument is set for observation by bringing the indices into contact with the mercury, by means of a small magnet, which attracts the steel through the glass, so that it is readily drawn up or down. They should be drawn nearly to the top of the limbs when it is desired to remove the instrument, which should be carefully carried in the vertical position; for should it be inverted, or laid flat, the spirit may get among the mercury, and so break up the column as to require the skill of a maker to put it in order again. For transmission by ordinary conveyances, it requires that attention be given to keep it vertical. The entanglement of a small portion of mercury with the indices is sometimes a source of annoyance in this instrument, for the readings are thereby rendered somewhat incorrect. Small breakages in the mercury, either from intervening bubbles of spirit or adhesion to the indices, may generally be rectified by cautiously tapping the frame of the instrument, so as to cause the mercury to unite by the assistance thus given to its superior gravity.

These thermometers, when carefully made and adjusted to a standard thermometer, are strongly recommended for ordinary purposes, where strict scientific accuracy is not required. This is also the only fluid thermometer applicable for determining the temperature of the sea at depths.

CHAPTER VIII.

RADIATION THERMOMETERS.

82. Solar and Terrestrial Radiation considered.—The surface of the earth absorbs the heat of the sun during the day, and radiates heat into space during the night. The envelope of gases and vapour, which we call the atmosphere, exerts highly important functions upon these processes. Thanks to the researches of Professor Tyndall, we are now enabled to understand these functions much more clearly than heretofore. His elaborate, patient, and remarkably sagacious series of experiments upon radiant heat, have satisfactorily demonstrated that *dry* air is as transparent to radiant heat as the vacuum itself; while air *perfectly saturated* with aqueous vapour absorbs more than five per cent. of radiant heat, estimated by the thermal unit adopted for the galvanometer indications of the effect upon a thermo-electric pile.

Aqueous vapour, in the form of fog or mist, as is well known, gives to our sensation a feeling of cold, and interferes with the healthy action of the skin and the lungs; the cause being its property of absorbing heat from our person.

Air containing moisture in an invisible state likewise exerts a remarkable influence in radiating and absorbing heat. By reason of these properties, aqueous vapour acts as a kind of blanket upon the ground, and maintains upon it a higher temperature than it would otherwise have. "Regarding the earth as a source of heat, no doubt at least ten per cent. of its heat is intercepted within ten feet of the surface." Thus vapour—whether transparent and invisible, or visible, as cloud, fog, or mist—is intimately connected with the important operations of solar and terrestrial radiation. Cloudy, or humid days, diminish the effect upon the soil of solar radiation ; similar nights retard the radiation from the earth. A dry atmosphere is the most favourable for the direct transmission of the sun's rays; and the withdrawal of the sun from any region over which the air is dry, must be followed by very rapid cooling of the soil. "The removal, for a single summer night, of the aqueous vapour from the atmosphere which covers England, would be attended by the destruction of every plant which a freezing temperature could kill. In Sahara, where ' the soil is fire and the wind is flame,' the refrigeration at night is often painful to bear. Ice has been formed in this region at night. In Australia, also, the *diurnal range* of temperature is very great, amounting, commonly, to between 40 and 50 degrees. In short, it may be safely predicted, that wherever the air is *dry*, the daily thermometric range will be great. This, however, is quite different from saying that when the air is *clear*, the thermometric range will be great. Great clearness to light is perfectly compatible with great opacity to heat; the atmosphere may be charged with aqueous vapour while a deep blue sky is overhead; and on such occasions the terrestrial radiation would, notwithstanding the 'clearness,' be intercepted." The great range of the thermometer is attributable to the absence of that protection against gain or loss of heat which is afforded when aqueous vapour is present in the air; and during such weather the rapid abstraction of moisture from the surface of plants and animals is very deleterious to their healthy condition. "The nipping of tender plants by frost, even when the air of the garden is some degrees above the freezing temperature, is also to be referred to chilling by radiation." Hence the practice of gardeners of spreading thin mats, of bad radiating material, over tender plants, is often attended with great benefit.

By means of the process of terrestrial radiation ice is artificially formed in Bengal, "where the substance is never formed naturally. Shallow pits are dug, which are partially filled with straw, and on the straw flat pans containing water which had been boiled is exposed to the clear firmament. The water is a very powerful radiant, and sends off its heat into space. The heat thus lost cannot be supplied from the earth—this source being cut off by the non-conducting straw. Before sunrise a cake of ice is formed in each vessel. . . . To produce the ice in abundance, the atmosphere must not only be clear, but it must be comparatively free from aqueous vapour."

Considering, therefore, the important consequences attending both terrestrial and solar radiation, it appears to us that observations from radiation thermometers are of much more utility in judging of climate than is usually supposed. These observations are very scanty; and what few are upon record are not very reliable, principally from bad exposure of the instruments, while the want of uniformity in construction may be another cause. Herschell's actinometer and Pouillet's pyrheliometer, instruments for ascertaining the absolute heating effect of the sun's rays, should, however, be more generally employed by meteorologists. In comparing observations on radiation it should be kept in mind, that " the difference between a thermometer which, properly confined [or shaded], gives the true temperature of the night air, and one which is permitted to radiate freely towards space, must be greater at high elevations than at low ones;"* because the higher the place, the less the thickness of the vapour-screen to intercept the radiation.

83. Solar Radiation Thermometer.—" As the interchange of heat between two bodies by radiation depends upon the relative temperature which they respectively possess, the earth, by the rays transmitted from the sun during the day, must be continually gaining an accession of heat, which would be far from being counterbalanced by the opposite effect of its own radiation into space. Hence, from sunrise till two or three hours after mid-day, the earth goes on gradually increasing in temperature, the augmentation being greatest where the surface consists of materials calculated, from their colour and texture, to absorb heat, and where it is deficient in

---

* The quotations in this section are from Tyndall's *Heat considered as a Mode of Motion.*

moisture, which, by its evaporation, would have a tendency to diminish it."*   It is, therefore, important to have instruments for measuring the efficacy of solar radiation, apart from those for exhibiting the temperature of the place in the shade.

Fig. 63 shows the arrangement of Negretti & Zambra's maximum thermometer,

Fig. 63.

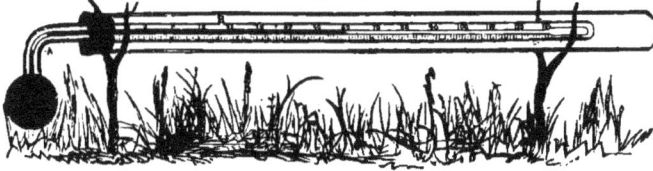

for registering the greatest heat of the sun's direct rays, hence called a *solar radiation thermometer*.   It has a blackened bulb, the scale divided on its own stem, and the divisions protected by a glass shield.   In use it should be placed nearly horizontally, resting on Y supports of wood or metal, with its bulb in the full rays of the sun, resting on grass, and, if possible, so that lateral winds should not strike the bulb ; and at a sufficient distance from any wall, so that it does not receive any *reflected* heat from the sun.   Some observers place the thermometer as much as two feet from the ground.   It would be very desirable if one uniform plan could be recognized : that of placing the instrument as indicated in the figure appears to be most generally adopted, and the least objectionable.

84. Vacuum Solar Radiation Thermometer.—In order that the heat absorbed by the blackened bulb of the solar radiation thermometer may not in part be carried off by the currents of air which would come into contact with it, the instrument has been improved by Messrs. Negretti and Zambra into the *vacuum solar radiation thermometer*, as illustrated by fig. 64.

Fig. 64.

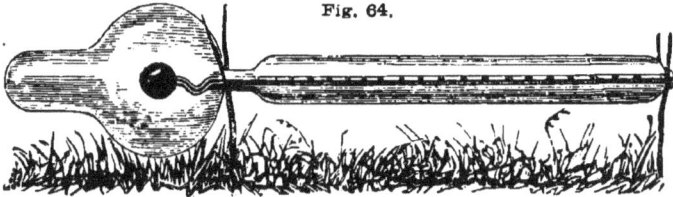

This consists of a blackened-bulb radiation thermometer, enclosed in a glass tube and globe, from which all air is exhausted.   Thus protected from the loss of heat which would ensue if the bulb were exposed, its indications are from 20° to 30° higher than when placed side by side with a similar instrument with the bulb exposed to the passing air.   At times when the air has been in rapid motion, the difference between the reading of a thermometer giving the true temperature of the

---

* Dr. Daubeny, F.R.S., *On Climate.*

air in the shade, and an ordinary solar radiation thermometer, has been 20° only, whilst the difference between the air temperature and the reading of a radiation thermometer in vacuo has been as large as 50°. It is also found that the readings are almost identical at distances from the earth varying from six inches to eighteen inches. By the use of this improvement, it is hoped that the amounts of solar radiation at different places may be rendered comparable; hitherto they have not been so; the results found at different places cannot be compared, as the bulbs of the thermometers are under very different circumstances as to exposure and currents of air. Important results are anticipated from this arrangement. The observations at different places are expected to present more agreement. Observers would do well to note carefully the effect of any remarkable degree of intensity in the solar heat upon particular plants, crops, fruit or other trees.

85. Terrestrial Radiation Thermometer is an alcohol minimum thermometer, with the graduations etched upon the stem, and protected by a glass shield, as shown in figure 65, instead of being mounted on a frame. The bulb is transparent; that

Fig. 65.

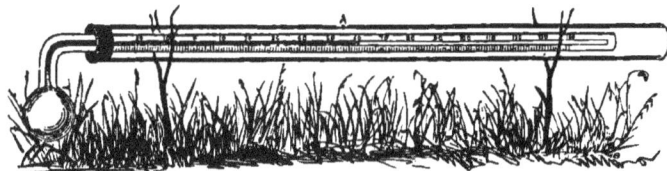

is to say, the spirit is not coloured.

In use, it should be placed with its bulb fully exposed to the sky, resting on grass, the stem being supported by little forks of wood. The precautions required with this thermometer are similar to those for ordinary spirit thermometers, explained at page 76.

86. Æthrioscope.—The celebrated experimental philosopher, Sir John Leslie, was the inventor of this instrument, the purpose of which is to give a comparative idea of the radiation proceeding from the surface of the earth towards the sky. It consists, as represented in fig. 66, of two glass bulbs united by a vertical glass tube, of so fine a bore that a little coloured liquid is supported in it by its own adhesion, there being air confined in each of the bulbs. The bulb, A, is enclosed in a highly polished brass sphere, D, made in halves and screwed together. The bulb, B, is blackened and placed in the centre of a metallic cup, C, which is well gilt on the inside, and which may be covered by a top, F. The brass coverings defend both bulbs from solar radiation, or any adventitious source of heat.

Fig. 63.

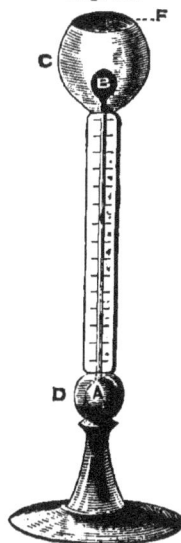

When the top is on, the liquid remains at zero of the scale. On removing the top and presenting the instrument to a clear sky, either by night or by day, the bulb, B, is cooled by terrestrial radiation, while the bulb, A, retains the temperature of the air. The air confined in B, therefore, contracts; and the elasticity of that within A forces the liquid up the tube, to a height proportionate to the intensity of the radiation. Such is the sensitiveness of the instrument, that the smallest cloud passing over it checks the rise of the liquid. Sir John Leslie says:—"Under a clear blue sky, the *æthrioscope* will sometimes indicate a cold of fifty millesimal degrees; yet, on other days, *when the air seems equally bright*, the effect is hardly 80°." This anomaly, according to Dr. Tyndall, is simply due to the difference in the quantity of aqueous vapour present in the atmosphere. The presence of invisible vapour intercepts the radiation from the æthrioscope, while its absence opens a door for the escape of this radiation into space.

87. Pouillet's Pyrheliometer.—"This instrument is composed of a shallow cylinder of steel, A, fig. 67, which is filled with mercury. Into the cylinder a

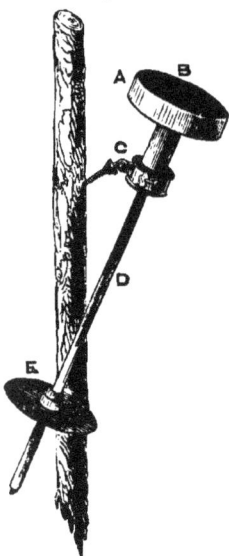

Fig. 67.

thermometer, D, is introduced, the stem of which is protected by a piece of brass tubing. We thus obtain the temperature of the mercury. The flat end of the cylinder is to be turned towards the sun, and the surface, B, thus presented is coated with lamp black. There is a collar and screw, C, by means of which the instrument may be attached to a stake driven into the ground, or into the snow, if the observations are made at considerable heights. It is necessary that the surface which receives the sun's rays should be perpendicular to the rays; and this is secured by appending to the brass tube which shields the stem of the thermometer, a disk, E, of precisely the same diameter as the steel cylinder. When the shadow of the cylinder accurately covers the disk, we are sure that the rays fall, as perpendiculars, on the upturned surface of the cylinder.

"The observations are made in the following manner:—First, the instrument is permitted, not to receive the sun's rays, but to radiate its own heat for five minutes against an unclouded part of the firmament; the decrease of the temperature of the mercury consequent on this radiation is then noted. Next, the instrument is turned towards the sun, so that the solar rays fall perpendicularly upon it for five minutes; the augmentation of heat is now noted. Finally, the instrument is turned again towards the firmament, away from the sun, and allowed to radiate for another five minutes, the sinking of the thermometer being noted as before. In order to obtain the whole heating power

of the sun, we must add to his observed heating power the quantity
lost during the time of exposure, and this quantity is the mean of the
first and last observations. Supposing the letter R to represent the
augmentation of temperature by five minutes' exposure to the sun,
and that t and t^1 represent the reductions of temperature observed
before and after, then the whole force of the sun, which we may call T,
would be thus expressed:—$T = R + \frac{1}{2} (t + t^1)$.

" The surface on which the sun's rays here fall is known; the quantity of mercury within the cylinder is also known; hence we can express
the effect of the sun's heat upon a given area, by stating that it is
competent, in five minutes, to raise so much mercury so many degrees
in temperature."—*Dr. Tyndall's " Heat considered as a Mode of
Motion."*

88. Sir John Herschell's Actinometer, for ascertaining the
absolute heating effect of the solar rays, in which *time* is considered
one of the elements of observation, is illustrated by fig. 68. The
actinometer consists of a large cylindrical thermometer bulb, with a
scale considerably lengthened, so that minute changes may be easily
seen. The bulb is of transparent glass filled with a deep blue liquid,
which is expanded when the rays of the sun fall direct on the bulb.
To take an observation, the actinometer is placed in the shade for one
minute and read off; it is then exposed for one minute to sunshine, and
its indication recorded; it is finally restored to the shade, and its
reading noted. The mean of the two readings in the shade, subtracted
from that in the sun, gives the actual amount of expansion of the
liquid produced by the sun's rays in one minute of time. For further
information, see *Report of the Royal Society on Physics and Meteorology;*
or *Kæmtz's Meteorology,* translated by C. V. Walker; or the *Admiralty
Manual of Scientific Instructions.*

Fig. 68.

CHAPTER IX.

DEEP-SEA THERMOMETERS.

89. On Sixe's Principle.—Thermometers for ascertaining the temperature of the sea at various depths are constructed to register either the maximum or minimum temperature, or both. The principle of each instrument is that of Sixe. There are very few parts of the ocean in which the temperature below is greater than at the surface, except in the Polar Seas, where it is generally found to be a few degrees warmer at considerable depths than at the surface. When the instrument is required to register only one temperature, it can be made narrower and more compact—a great advantage in sounding; and with less length of bulb and glass tube, so that the liability of error is diminished. Hence, the minimum is the most generally useful for deep-sea soundings. These thermometers must be sufficiently strong to withstand the pressure of the ocean at two or three miles of depth, where there may be a force exerted to compress them exceeding three or four hundred atmospheres (of 15 lbs. to the square inch).

Many have been the contrivances for obtaining correct deep-sea indications. Thermometers and machines of various sorts have been suggested, adopted, and eventually abandoned as only approximate instruments. The principal reason for such instruments failing to give correct or reliable indications, has been that the weight or pressure on the bulbs at great depths has interfered with the correct reading of the instruments. Thermometers have been enclosed in strong water-tight cases to resist the pressure; but this contrivance has only had the tendency to retard the action, so much so as to throw a doubt on the indications obtained by the instrument so constructed.

The thermometers constructed by Messrs. Negretti and Zambra for this purpose do not differ materially from those usually made under the denomination of Sixe's thermometers, except in the following most important particular:—The usual Sixe's thermometers have a central reservoir or cylinder containing alcohol; this reservoir, which is the only portion of the instrument likely to be affected by pressure, has been, in Negretti and Zambra's new instrument, superseded by a strong outer cylinder of glass, containing mercury and rarefied air; by this means the portion of the instrument susceptible of compression, has been so strengthened that no amount of pressure can possibly make the instrument vary. This instrument has been tested in every possible manner, and the results have been highly satisfactory, so much so as to place their reliability beyond any possible doubt.

The scales are made of porcelain, and are firmly secured to a back of oak, which holds in a recess the bulb with its protecting shield, and is rounded off so as to fit easily and firmly in a stout cylindrical copper case, in which the thermometer is sent down when sounding (see fig. 69). The lid of the case is made to fit down closely, and water-tight. At the bottom of the case is a valve opening upward; and the lid has a similar valve. These allow the water to pass through the case as the instrument sinks, so that the least amount of obstruction is offered to the descent. At the lower end of the case is a stout brass spring, to protect the instrument from a sudden jar if it should touch the bottom while descending rapidly. As the instrument is drawn up, the valves close with the weight of water upon them, and it arrives at the surface filled with water brought up from its lowest position. The deep-sea thermometers used in the Royal Navy are of this pattern.

90. Johnson's Metallic Deep-Sea Thermometer.—The objection to the employment of mercurial thermometers for ascertaining the temperature of the ocean at depths, arising from the compression of the bulbs, which was of such serious consequence previous to the modification made in the construction of the instrument by Messrs. Negretti and Zambra, led to the construction of a metallic thermometer altogether free from liability of disturbance from compression by the surrounding water; which, however, is certainly not so sensitive to changes of temperature as mercury. This instrument is the invention of Henry Johnson, Esq., F.R.A.S., and is thus described by him:—

"During the year 1844 some experiments were made by James Glaisher, Esq., F.R.S., on the temperature of the water of the Thames near Greenwich at the different seasons of the year; when that gentleman found that the indications of temperature were greatly affected by the pressure on the bulbs of the thermometers. At a depth of 25 feet this pressure would be nearly equal to the presence of three-fourths of an atmosphere. These observations demonstrate the importance of using in deep-sea soundings an instrument free from liability of disturbance from compression by the surrounding water, and have ultimately led to the construction of the thermometer now to be described.

"The instrument is composed of solid metals of considerable specific gravity, viz. of brass and steel, the specific gravity of these metals being 8·39 and 7·81 respectively. They are therefore not liable to compression by the water, which under a pressure of 1,120 atmospheres, or at a depth of 5,000 fathoms in round numbers, acquires a density or specific gravity of 1·06. In the construction of this instrument, advantage has been taken of the well-known difference in the ratios of expansion and contraction by heat and cold of brass and steel, to form compound bars of thin bars of these metals riveted together; and which will be found to assume a slight curve in one direction when heat has expanded the brass more than the steel, and a slight one in the contrary direction when cold has contracted the brass more than the steel.

"The indications of the instrument record the motions under changes of temperature of such compound bars; in which the proportion of brass, the more dilatable metal, is two-thirds, and of steel one-third.

Fig. 70.

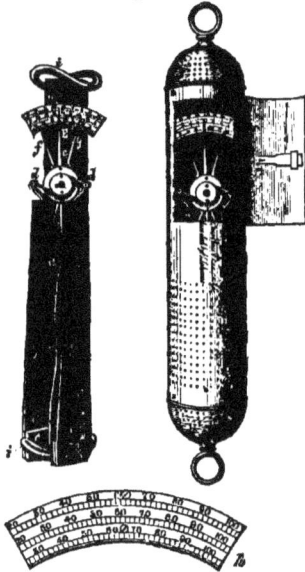

"Upon one end of a narrow plate of metal about a foot long, *a*, are fixed three scales of temperature, *h*, which ascend from 25° to 100° F., and which are shown more clearly in the drawing detached from the instrument. Upon one of these scales the present temperature is shown by the pointer, *e*, which turns upon a pivot in its centre. The register index, *g*, to the maximum temperature, and the index, *f*, to the minimum temperature, are moved along the other scales by the pin upon the moving pointer, at *e*, where they are retained by stiff friction. At equal distances from the centre of the pointer are two connecting pieces, *d d*, by which it is attached to the free ends of two compound bars, *b b*, and its movements correspond with the movements of the compound bars under variations of temperature. The other ends of the bars are fastened by the plate, *c*, to the plate, *a*, on which the scales of temperature are fixed. The connection of the bars with both sides of the centre of the pointer prevents disturbance of indication by lateral concussion. The case of the instrument has been improved at the suggestion of Admiral FitzRoy, and now presents to the water a smooth cylindrical surface, with rounded ends, and without projection of fastenings.

"In surveying expeditions, this instrument would be found useful in giving notice of variation of depth of water, and of the necessity for taking soundings. A diminution of the temperature of water has been observed by scientific voyagers to accompany diminution of depth, as on nearing land, or approaching hidden rocks or shoals. Attention would also thus be attracted to the vicinity of icebergs."

This thermometer might easily be modified to serve for several other important purposes, such as the determination of the temperature of intermittent hot springs, and mud volcanoes.

The principle of this thermometer is not altogether new; but the duplicate arrangement of the bars, which effectually prevents the movement of the indices by any shaking, and the application are certainly novel. Professor Trail, in the *Library of Useful Knowledge*, writes :—"In 1803, Mr. James Crighton, of Glasgow, published a new 'metallic thermometer,' in which the unequal expansion of zinc and iron is the moving power. A bar is formed by uniting a plate of zinc (fig. 71), *c d*, 8 inches long, 1 inch broad, and ¼ inch thick, to a plate of iron, *a b*, of the same length.   The lower extremity of the compound bar is firmly attached to a mahogany board at *e e;* a pin, *f*, fixed to its upper end, plays in the forked opening in the short arm of the index, *g*. When the temperature is raised, the superior expansion of the zinc, *c d*, will bend the whole bar, as in the figure; and the index, *g*, will move along the graduated arc, from right to left, in proportion to the temperature.   In order to convert it into a *register thermometer*, Crighton applied two slender hands, *h h*, on the axis of the index; these lie below the index, and are pushed in opposite directions by the stud, *i*,—a contrivance seemingly borrowed from the instrument of Fitzgerald," a complicated metallic thermometer, described by the Professor previously.

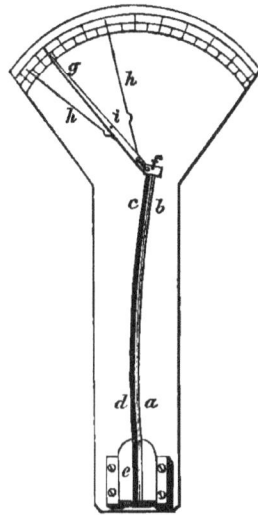

Fig. 71.

CHAPTER X

91. Ebullition.—The temperature at which a fluid *boils* is called the *boiling-point* of that particular fluid. It is different for different liquids; and, moreover, in the same liquid it varies with certain changes of circumstance. Thus the same liquid in various states of purity would have its boiling temperature altered in a slight degree. There is also an intimate connection with the pressure under which a fluid is boiled, and its temperature of ebullition. Liquids boiled in the open air are subjected to the atmospheric pressure, which is well known to vary at different times and places; and the boiling-point of the liquid exhibits corresponding changes. When the pressure is increased on the surface of any fluid, the temperature of ebullition rises; and with a decrease of pressure, the boiling goes on at a lower degree of heat.

In the case of water, we commonly state the boiling-point to be 212° F.; but it is only so at the level of the sea, under the mean pressure of the atmosphere, represented, in the latitude of London, by a column of 29·905 inches of mercury, at a temperature of 32° F., and when the water is fresh and does not contain any matter chemically dissolved in it. When steam is generated and confined in a boiler, the pressure upon the boiling water may be several times greater than that of the atmosphere. Experimentally it has been found, that if the pressure in the boiler be 25 ℔s. on the square inch, the temperature of the boiling water, and of the steam likewise, is raised to 241°; while under the exhausted receiver of an air-pump, water will boil at 185°, when the pressure is reduced to 17 inches of mercury.

92. Relation between the Boiling-Point and Elevation.—Now, as the atmospheric pressure is diminished by ascent, as shown by the fall of mercury in the barometer, it follows that in elevated localities water, or any other fluid, heated in the open air, will boil at a temperature lower than at the sea-level. Therefore, there must be some relation between the height of a hill, or mountain, and the temperature at which a fluid will boil at that height. Hence, the thermometer, as used to determine the boiling-point of fluids, is also an indicator of the atmospheric pressure; and may be used as a substitute for the barometer in measuring elevations.

If the atmospheric pressure were constant at the sea-level, and always the same for definite heights, we might expect the boiling-points of fluids also to be in exact accordance with height; and the relation once ascertained, we could readily, by means of the thermometer and boiling water, determine an unknown height, or for a known elevation assert the boiling temperature of a liquid. However, as the atmospheric pressure is perpetually varying at the same place, within certain limits, so there are, as it were, sympathetic changes in the boiling temperatures of fluids.

It follows from this, that heights can never be accurately measured, either by the barometer or the boiling-point thermometer, by simply observing at the places whose elevations are required. To determine a height with any approach to accuracy, it is necessary that a similar observation should be made at the same time at a lower station, not very remote laterally from the upper, and that they should be many times repeated. When such observations have been very carefully conducted, the height of the upper station above the lower may be ascertained with great precision, as has been repeatedly verified by subsequent trigonometrical measurement of elevations so determined. If the lower station be at the sea-level, of course the absolute height of the upper is at once obtained.

93. Mountain Thermometer; sometimes called Hypsometric Apparatus.— We have now to examine the construction of the boiling-point thermometer, and its necessary appendages, as adapted for the determination of heights.

Messrs. Negretti and Zambra's arrangement of the instrument is shown in figures 72 and 73.

Fig. 72.

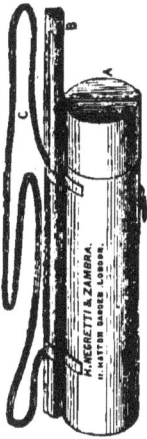

Fig. 73.

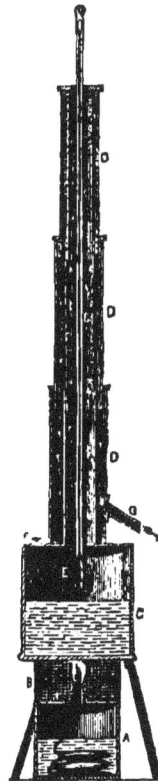

The thermometer is made with an elongated bulb, so as to be as sensitive as possible. The scale, about a foot long, is graduated on the stem, and ranges from 180° to 214°, each degree being sufficiently large to show the divisions of tenths of a degree. A sliding metallic vernier might perhaps with advantage be attached to the stem, which would enable the observer to mark hundredths of a degree; which, however, he can pretty well do by estimation. The boiler is so contrived as to allow, not only the bulb, but the stem also of the thermometer, to be surrounded by the steam. The arrangement is readily understood by reference to the accompanying diagram, fig. 73.

C, is a copper boiler, supported by a tripod stand so as to allow a spirit-lamp, A, made of metal to be placed underneath. The flame from the lamp may be surrounded by a fine wire gauze, B, which will prevent it being extinguished when experimenting in the external air. $E E E$, is a three-drawn telescope tube, proceeding from the boiler, and open also at top. Another tube, similarly constructed, envelops this, as shown by $D D D$. This tube is screwed to the top of the boiler, and has two openings, one at the top to admit the thermometer, the other low down, G, to give vent to the steam. As the steam is generated, it rises in the inner tube, passes down between the tubes, and flows away at G. The thermometer is passed down, supported by an india-rubber washer, fitting steam tight, so as to leave the top of the mercury, when the boiling-point is attained, sufficiently visible to make the observation. The telescopic movement, and the mode of supporting the thermometer, enable the observer always to keep the bulb near the water, and the double tube gives all the protection required to obtain a steady boiling-point. Some boiling-point thermometers are constructed with their scales altogether exposed to the air, which may be very cold, and consequently may contract to some extent the thread of mercury outside the boiler. The steam, having the same temperature as the boiling water, keeps the tube, throughout nearly its whole length, at the same degree of heat, in the apparatus described. The whole can be packed in a tin case very compactly and securely for travelling, as in fig. 72.

Directions for Using.—When the apparatus is required for practical use, sufficient water must be poured into the boiler to fill it about one third, through an opening, F, which must be afterwards closed by the screw plug. Then apply the lighted lamp. In a short time steam will issue from G; and the mercury in the thermometer, kept carefully immersed, will rise rapidly until it attains a stationary point, which is the boiling temperature. The observation should now be taken and recorded with as much accuracy as possible, and the temperature of the external air must be noted at the same time by an ordinary thermometer.

The water employed should be pure. Distilled water would therefore be the best. If a substance is held mechanically suspended in water, it will not affect the boiling-point. Thus, muddy water would serve equally as well as distilled water. However, as it cannot be readily ascertained that nothing is dissolved chemically when water is dirty, we are only correct when we employ pure water.

94. Precautions to ensure correct Graduation.—Those who possess a boiling-point thermometer should satisfy themselves that it has been correctly graduated. To do this, it is advisable to verify it with the reading of a standard barometer reduced to 32° F. The table of "Vapour Tension" (given at p. 62) will furnish the means of comparison. Thus, if the reduced reading of the barometer, corrected also for latitude, be 29·922, the thermometer should show 212° as the boiling-point of water at the same time and place; if 29·745, the thermometer should read 211·7; and so on as per table. In this way the error of the chief point of the scale can be obtained. Other parts of the scale may be checked with a standard thermometer, by subjecting both to the same temperature, and comparing their indications. The graduations as fixed by some makers are not always to be trusted; and this essential test should be conducted with the utmost nicety and care.

Admiral FitzRoy writes, in his *Notes on Meteorology:*—"Each degree of the boiling-point thermometer is equivalent to about 550 *feet of ascent*, or one-tenth to

55 feet; therefore, the smallest error in the graduation of the thermometer itself will affect the height deduced materially.

"In the thermometer which is graduated from 212° (the boiling-point) to 180°, similarly to those intended for the purpose of measuring heights, there must have been a starting point, or zero, from which to begin the graduation. I have asked an optician in London how he fixed that zero, the boiling-point. 'By boiling water at my house,' he replied. 'Where is your house?' In such a part of the town, he answered. I said: 'What height is it above the sea?' to which he replied, 'I do not know;' and when I asked the state of the barometer when he boiled the water, whether the mercury was high or low, he said that he had not looked at it! Now, as this instrument is intended to measure heights and to decide differences of some hundred, if not thousand feet upwards, at least one should endeavour to ascertain a reliable starting point. From inquiries made, I believe that the determination of the boiling-point of ordinary thermometers has been very vague, not only from the extreme difficulties of the process itself (which are well known to opticians), but from the radical errors of not allowing for the pressure of the atmosphere at the time of graduation—which may be much, even an inch higher or lower, than the mean, or any *given height*—while the elevation of the place above the level of the sea is also unnoticed. Then there is another source of error, a minor one, perhaps: the inner limit, the 180° point, is fixed only by comparison with another thermometer; it may be right, or it may be very much out, as may be the intermediate divisions; for the difficulty of ascertaining degree by degree is great: and it must be remembered that the measurement of a very high mountain depends upon those inner degrees from 200° down to 180°, thereabouts. Hence, the difficulty of making a reliable observation by boiling water seems to be greater than has been generally admitted."

95. Method of Calculating Heights from Observations with the Mountain Thermometer.—Having considered how to make observations with the proper care and accuracy, it becomes necessary to know how to deduce the height by calculation. That a constant intimate relation exists between the boiling temperature of water and the pressure of the air, we have already learned. This knowledge is the result of elaborate experiments made by several scientific experimentalists, who have likewise constructed formulæ and tables for the conversion of the boiling temperatures into the corresponding pressures of vapour, or, which is equivalent, of the atmosphere, when the operation is performed in the open air. As might be expected, there is not a perfect accord in the results arrived at by different persons. Regnault is the most recent, and his experiments are considered the most reliable.

From Regnault's table of vapour tension, we can obtain the pressure in inches of mercury at 32°, which corresponds to the observed boiling-point; or *vice versa*, if required. From the pressure, the height may be deduced by the method for finding heights by means of the barometer.

H

The following table expresses very nearly the elevation in feet corresponding to a fall of 1° in the temperature of boiling water:—

Boiling Temperatures between.		Elevation in Feet for each Degree.
214° and 210 —	....	520
210 and 200—	....	530
200 and 190	....	550
190 and 180	....	570

These numbers agree very well with the results of theory and actual observation. The assumption is that the boiling-point will be diminished 1° for each 520 feet of ascent until the temperature becomes 210°, then 530 feet of elevation will lower it one degree until the water boils at 200°, and so on; the air being at 32°.

Let H represent the vertical height in feet between two stations; B and b, the boiling-points of water at the lower and upper stations respectively; f, the factor found in the above table. Then

$$H = f\,(B - b)$$

Further, let m be the mean temperature of the stratum of air between the stations. Now, if the mean temperature is less than 32°, the column of air will be shorter; and if greater, longer than at 32°. According to Regnault, air expands $\frac{1}{491.13}$ or ·002036 of its volume at 32°, for each degree increase of heat. Calling the correction due to the mean temperature of air C, its value will be found from the equation,

$$C = H\,(m - 32)\ \cdot 002036$$

Calling the corrected height H', it will be found from the formula,

$$H' = H + H\,(m - 32)\ \cdot 002036$$

that is, $$H' = H\left\{1 + (m - 32)\ \cdot 002036\right\}$$

and substituting the value of H,

$$H' = f\,(B - b)\left\{1 + (m - 32)\ \cdot 002036\right\}$$

Strictly, according to theoretical considerations, there is a correction due to latitude, as in the determination of heights by the barometer; but its value is so small that it is practically of no importance.

If a barometer be observed at one of the stations, the table of vapour tensions (p. 62) will be useful in converting the pressure into the corresponding boiling-point, or vice versa; so that the difference of height may be found either by the methods employed for the boiling-point thermometer or the barometer.

In conclusion, it may be remarked that observers who have good instruments at considerable elevations, as sites on mountains or plateaus, would confer a benefit to science, by registering for a length of time the barometer along with the boiling temperature of water, as accurately as possible. Such observations would serve to verify the accuracy of theoretical deductions, and fix with certainty the theoretical scale with the barometer indications.

Example, in calculating Heights from the Observations of the Boiling-point of Water.—1. At Geneva the observed boiling-point of water was 209°·335; on the Great St. Bernard it was 197°·64; the mean temperature of the intermediate air was 63°·5; required the height of the Great St. Bernard above Geneva.

Method by formula :—

$$H' = f\,(B - b)\,\left\{\,1 + (m - 32°)\,\cdot002036\,\right\}$$

In this case f is between 530 and 550, or 540.

$B = 209\cdot335$	$m = 63\cdot5$
$b\ = 197\cdot64$	32
$\overline{11\cdot695}$	$\overline{31\cdot5}$
$f\ =\ \ \ 540$	·002036
$\overline{6315\cdot3}$	$\overline{0\cdot0641340}$
$1\cdot064$	1
$H' = 6719\cdot5$ feet.	$1\cdot064$

Method by Tables supplied with boiling-point apparatus made by Messrs. Negretti and Zambra :—

209·335	gives	1464	in Table I.
197·64	,,	7736	,,
		$\overline{6272}$	
63·5	,,	1·07	in Table II.
Height	...	$\overline{6711}$	

96. Thermometers for Engineers,—1*st.* *Salinometer.*— Under the circumstances at which fresh water boils at 212°, sea water boils at 213°·2. The boiling temperature is raised by the chemical solution of any substance in the water, and the more with the increase of matter dissolved.

From a knowledge of this principle, marine engineers make use of the thermometer to determine the amount of salts held in solution by the water in the boilers of sea-going steamers. Common sea-water contains $\frac{1}{33}$ of its volume of salt and other earthy matters. As evaporation proceeds, the solution becomes proportionally stronger, and more heat is required to produce steam. The following table from the work of Messrs. Main and Brown, on the Marine Steam-Engine, shows the relation between the boiling-point under the mean pressure of the atmosphere, or 30 inches of mercury, and the proportion of matter dissolved in the water :—

Proportion of Salt in 100 parts of water 0 ...... Boiling-point 212°

,,	,,	$\frac{1}{33}$	......	,,	213·2
,,	,,	$\frac{2}{33}$	......	,,	214·4
,,	,,	$\frac{3}{33}$	......	,,	215·5
,,	,,	$\frac{4}{33}$	......	,, ·	216·6
,,	,,	$\frac{5}{33}$	......	,,	217·9
,,	,,	$\frac{6}{33}$	......	,,	219·0
,,	,,	$\frac{7}{33}$	......	,,	220·2
,,	,,	$\frac{8}{33}$	......	,,	221·4
,,	,,	$\frac{9}{33}$	......	,,	222·5
,,	,,	$\frac{10}{33}$	......	,,	223·7
,,	,,	$1\frac{1}{33}$	......	,,	224·9
,,	,,	$1\frac{2}{33}$	......	,,	226·0

When the salts in solution amount to $1\frac{2}{33}$, the water is saturated. It has also been ascertained that, when a solution of $\frac{3}{33}$ is attained, incrustation of the substances commences on the boiler. Hence, it is a rule with engineers to expel some of the boiling water, when the thermometer indicates a temperature of 216°, and introduce some more cold water, in order to prevent incrustation, which not only injures the boiler, but opposes the passage of heat to the water. The thermometer used for this purpose should be very accurately graduated, and the scale must be considerably higher than, though it need not read much below 212°.

Fig. 74.

2nd. *Pressure Gauge.*—The elasticity of gases augments by increase of temperature, and *vice versa;* it follows, therefore, that when steam is generated in a closed boiler, its temperature rises beyond the boiling temperature of 212°, owing to the increased pressure upon the water. The law connecting the pressure and the corresponding temperature of steam is the same as that upon which the boiling of fluids under diminished atmospheric pressure takes place. Hence, the indications of the thermometer become exponents of steam pressure. Engineers are furnished, in works on the steam-engine, with tables, from which the pressure corresponding to a given temperature, or the converse, can be obtained by mere inspection.

Fig. 74 represents the thermometer employed as a steam-pressure gauge. It is fitted in a brass case, with screw-plug and washers for closing the boiler when the thermometer is not in use. The scale shows the pressure corresponding to the temperature, from 15 to 120 ℔s., above the atmospheric pressure, which is usually taken as 15 ℔s. on the square inch.

CHAPTER XI.

INSTRUMENTS FOR ASCERTAINING THE HUMIDITY OF THE AIR.

97. Hygrometric Substances.—The instruments devised for the purpose of ascertaining the humidity of the atmosphere are termed *hygrometers*. The earliest invented hygrometers were constructed of substances readily acted upon by the vapour in the air, such as hair, grass, seaweed, catgut, &c., which all absorb moisture, and thereby increase in length, and when deprived of it by drying they contract. Toy-like hygrometers, upon the principle of absorption, are still common as ornaments for mantel-pieces. A useful little instrument of this class, formed from the beard of the wild oat, is made to resemble a watch in external appearance, and is designed to prove the dampness or dryness of beds : a moveable hand points out on the dial the hygrometric condition of the clothes upon which the instrument is laid.

98. Saussure's Hygrometer, formerly used as a meteorologic **Fig. 75.** instrument, but now regarded as an ornamental curiosity, is represented in fig. 75. Its action depends upon a prepared hair, fixed at one end to the frame of the instrument, and wound round a pulley at the other. The pulley carries a pointer which has a counterpoise sufficient to keep the hair stretched. By this means the shrinking and lengthening of the hair cause the pointer to traverse a graduated arc indicating the relative humidity.

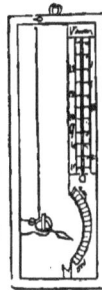

Such instruments, however ingenious, are not of scientific value; because they do not admit of rigid comparison, are liable to alter in their contractile and expansive properties, and cannot be made to indicate precisely alike.

99. Dew-Point.—The amount of water which the air can sustain in an invisible form increases with the temperature; but for every definite temperature there is a limit to the amount of vapour which can be thus diffused. When the air is cooled, the vapour present may be more than it can sustain; part will then be condensed as dew, rain, hail or snow, according to the meteorologic circumstances. The temperature which the air has when it is so fully saturated with vapour that any excess will be deposited as dew, is called the *dew-point*.

100. Drosometer.—" To measure the quantity of dew deposited each night, an instrument is used called a *Drosometer*. The most simple process consists in exposing to the open air bodies whose exact weight is known, and then weighing them afresh after they are covered with dew. According to Wells, locks of wool, weighing about eight grains, are to be preferred, which are to be divided [formed] into spherical masses of the diameter of about two inches."—*Kamtz.*

101. Humidity.—The proportion existing between the amount of vapour actually present in the air at any time, and the quantity necessary to completely saturate it, is called *the degree of humidity*. It is usually expressed in a centesimal scale, 0 being perfect dryness, and 100 complete saturation.

The pressure, or tension, of vapour at the dew-point temperature, divided by the tension of vapour at the air temperature and the quotient multiplied by 100, gives the degree of humidity. (Regnault's Tables should be used.)

Hence the utility of instruments for determining the dew-point.

102. Leslie's Hygrometer.—This instrument consists of a glass syphon tube, terminated with a bulb or ball at each end, turned outwards from each other, as in

Fig. 76.

fig. 76. The tube is partly filled with concentrated sulphuric acid, tinged by carmine. One of the balls is covered smoothly with fine muslin, and is kept continually moistened with pure water, drawn from a vase placed near it by the capillary attraction of a few strands of clean cotton-wick. The descent of the coloured liquid in the other stem will mark the diminution of temperature caused by the evaporation of the water from the humid surface. The drier the ambient air is, the more rapidly will the evaporation go on; and the cold produced will be greater. When the air is nearly saturated with moisture, the evaporation goes on slowly; the cold produced is moderate, because the ball regains a large portion of its lost heat from surrounding bodies; and the degree of refrigeration of the ball is an index of the dryness of the air.

" Should the water become frozen on the ball, this hygrometer will still act; for evaporation goes on from the surface of ice in proportion to the dryness of the air. Leslie estimates, that when the ball is moist, air, at the temperature of the ball, will take up moisture equal to the sixteen-thousandth part of its weight, for each degree of his hygrometer; and as ice in melting requires one-seventh of the caloric consumed in converting water into vapour, when the ball is frozen, the hygrometer will sink more than when wet by 1° in 7°; and hence, in the frozen state, we must increase the value of the degrees one-seventh: so that each of them will correspond to an absorption of moisture equal to one-fourteen-thousandth part of the weight of the air.

" When this hygrometer stands at 15°, the air feels damp; from 30° to 40°, we reckon it dry; from 50° to 60°, very dry; and from 70° upwards, we should call it intensely dry. A room would feel uncomfortable, and would probably be unwholesome, if the instrument in it did not reach 30°.* In thick fogs it keeps almost at the beginning of the scale. In winter, in our climate, it ranges from 5° to

---

* Leslie *On the Relations of Air, Heat, and Moisture.*

15°; in summer often from 15° to 55°; and sometimes attains 80° or 90°. The greatest degree of dryness ever noticed by Leslie was at Paris, in the month of September, when the hygrometer indicated 120°."—*Professor Trail, in " Library of Useful Knowledge."*

In estimating the value of the indications of this hygrometer, it should be borne in mind that the scale adopted by Leslie was *millesimal*, that is to say, from the freezing to the boiling-point of water was divided into a thousand parts; ten millesimal degrees are therefore equal to one of the scale of Celsius.

103. DANIEL'S HYGROMETER.

This instrument was invented about the year 1820, by Professor Daniel, the distinguished author of *Meteorological Essays;* and it entirely superseded all hygrometers depending upon the absorption of moisture. The form of the instrument is shown in fig. 77.

It consists of a glass tube, about one-eighth of an inch in diameter of bore, bent twice at right angles, and terminated, at each end, in a bulb about one inch and a quarter in diameter. In one limb of the tube is enclosed a delicate thermometer, which descends to the centre of the adjoining bulb, which is about three-parts filled with sulphuric ether. All the other parts of the tube are carefully freed from air, so that they are occupied by the vapour of the ether. This bulb is generally made of black glass; the other is transparent, but covered with a piece of fine muslin. The support for the tube has a thermometer attached, which shows the temperature of the external air. The tube can be removed from the stand, and the parts are made to pack, with a necessary phial of ether, in a small box, which can easily be got into the pocket.

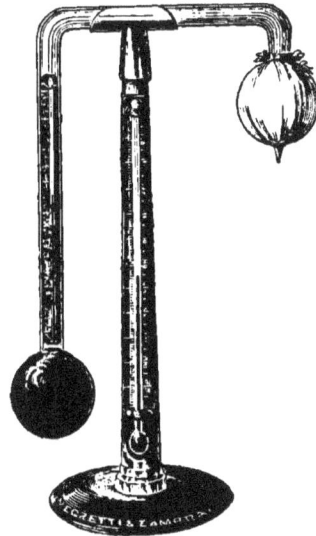

Fig. 77.

How to use the Hygrometer.—This instrument gives the dew-point by direct observation, which must be made in the following manner:—Having fixed the tube upon the stand, with the bulbs vertically downward, the ether is all caused to flow into the lower ball by inclining the tube. The temperature of the air is noted by the exposed thermometer. Then some ether is poured, from a dropping tube fitting into the neck of the phial, upon the muslin-covered bulb. The rapid evaporation of this ether cools the bulb and causes condensation of the ethereal vapour in its interior. This gives rise to rapid evaporation of the ether in the lower bulb, whereby its temperature is greatly reduced. The air in the vicinity is deprived of its warmth by the cold bulb, and is soon cooled to the temperature at which it is perfectly saturated with the vapour which it contains. Cooled ever so little below this temperature, some aqueous vapour will be condensed, and will form a dew upon the black-glass bulb. At the first indication of the deposit of dew the reading of the internal thermometer is taken : which is the dew-point.

This hygrometer has undeniable disadvantages. The surface upon which the dew condenses is small, and requires a peculiar direction of light in which to see it well. The observer, having his attention on the bulb and the thermometer, cannot always fix with precision the dew-point; and hence he is recommended to note the temperature at the appearance and at the disappearance of the dew, in order that the chance of error may be diminished. Without doubt, the necessarily long continuance of the observer near the instrument influences, to some extent, the observed temperatures; and the difficulty of not being always able to procure pure ether for the experiments is not the least of the drawbacks to the use of the instrument. Some of these disadvantages are obviated in Regnault's hygrometer.

Fig. 78.

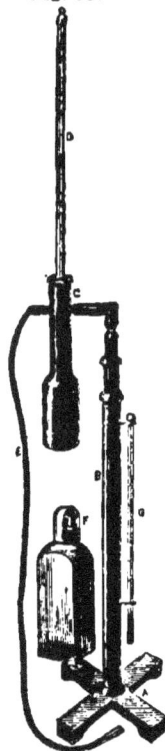

104.   REGNAULT'S CONDENSER HYGROMETER

(Fig. 78) consists of a tube, C, made of silver, very thin, and perfectly polished; the tube is larger at one end than the other, the large part being 1·8 inches in depth, by 0·8 in diameter; this is fitted tightly to a brass stand, B, with a telescopic arrangement for adjusting when making an observation.

The tube, C, has a small lateral tubulure, to which is attached an India-rubber tube, with ivory mouth-piece; this tubulure enters C at right angles near the top, and traverses it to the bottom of the largest part.

A delicate thermometer, D, is inserted through a cork, or India-rubber washer, at the open end of the tube, C, the bulb of which descends to the centre of its largest part.

G is an attached thermometer for taking the temperature of the air, and F is a bottle containing ether.

To use the Condenser Hygrometer, a sufficient quantity of ether is poured into the silver tube to cover the thermometer bulb : on allowing air to pass bubble by bubble through the ether, by breathing in the tube, E, an uniform temperature will be obtained; if the ether continues to be agitated, by breathing briskly through the tube a rapid reduction of temperature will be the result; at the moment the ether is cooled down to the dew-point temperature, the external surface of that portion of the silver tube containing ether will become covered with a coating of moisture, and the degree shown by the thermometer at that instant will be the temperature of the dew-point.

This form of hygrometer, for ascertaining by direct observation the dew-point, is so superior to Daniell's, both from its being more certain in its indications and economical in use, that Messrs. Negretti and Zambra have been induced to modify it, and reduce its price to little more than that of a good Daniell's Hygrometer.

105. Temperature of Evaporation.—When the air is not saturated with vapour, evaporation is going on with more or less activity, according as the temperature is high or low, rising or falling. Now vapour cannot be formed without an expenditure of heat; as we invariably find that the process of evaporation lowers the temperature of the liquid from which the vapour is produced, and, by communication, that of contiguous substances also. Thus the emigrant, crossing the line under the scorching influence of the vertical sun, wraps a wet towel round his can of water, swings it in the breeze, to evaporate the moisture of the towel, and obtains a glass of cool water. So also, European residents in India, during the hot season, spread out mats in their apartments, and keep them wet, in order that the evaporation may cool the air. This principle has been applied, for the purpose of ascertaining the hygrometric condition of the air, in the instrument known as Mason's hygrometer, or psychrometer, which is now in general use, from its simplicity, accuracy, and ease of observing.

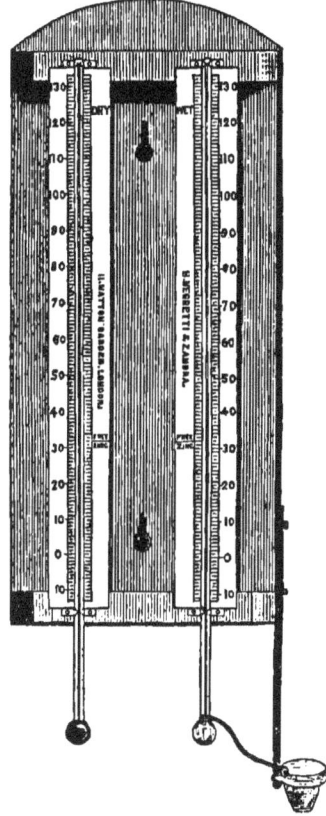

Fig. 79.

106. MASON'S HYGROMETER.

The Dry and Wet Bulb Hygrometer, or Psychrometer, known also as Mason's hygrometer (fig. 79), consists of two parallel thermometers, as nearly identical as possible, mounted on a wooden bracket, one marked *dry*, the other *wet*. The bulb of the wet thermometer is covered with thin muslin, and round the neck is twisted a conducting thread of lamp-wick, which passes into a vessel of water, placed at such a distance as to allow a length of conducting thread, of about three inches; the cup or glass is placed on one side, and a little beneath, so that the water within may not affect the reading of the *dry bulb thermometer*. In observing, the eye should be placed on a level with the top of the mercury in the tube, and the observer should refrain from breathing whilst taking an observation.

The *dry* bulb thermometer indicates the temperature of the air itself; while the *wet* bulb, cooled by evaporation, shows a lower temperature according to the rapidity of evaporation.

To find the Dew-point.—From the readings of the two thermometers, the dew-point can be deduced by formulæ (that known as Apjohn's is considered the most theoretically true), or from the valuable Hygrometric Tables by J. Glaisher, Esq., F.R.S.

For practical purposes in estimating the comparative humidity, the annexed table, which is a reduction from Mr. Glaisher's elaborate work, will be sufficient; it will at least serve to assist in familiarising the inexperienced in the value of the psychrometer's indications :—

Temperature by the Dry Bulb Thermometer.	Difference between Dry-bulb and Wet-bulb Readings.					
	2°	4°	6°	8°	10°	12°
	Degree of Humidity.					
34°	79	63	50	...	...	...
36	82	66	53	...	...	...
38	83	68	56	45	...	...
40	84	70	58	47	...	...
42	84	71	59	49	...	...
44	85	72	60	50	...	...
46	86	73	61	51	...	...
48	86	73	62	52	44	...
50	86	74	63	53	45	...
52	86	74	64	54	46	...
54	86	74	64	55	47	...
56	87	75	65	56	48	...
58	87	76	66	57	49	...
60	88	76	66	58	50	43
62	88	77	67	58	50	44
64	88	77	67	59	51	45
66	88	78	68	60	52	45
68	88	78	68	60	52	46
70	88	78	69	61	53	47
72	89	79	69	61	54	48
74	89	79	70	62	55	48
76	89	79	71	63	55	49
78	89	79	71	63	56	50
80	90	80	71	63	56	50
82	90	80	72	64	57	51
84	90	80	72	64	57	51
86	90	80	72	64	58	52

The total quantity of aqueous vapour which at any temperature can be diffused in the air being represented by 100, the per-centage of vapour actually present will be found in the table opposite the temperature of the dry thermometer, and under the difference between the dry-bulb and wet-bulb temperatures. The degree of humidity for intermediate temperatures and differences to those given in the table can be easily estimated sufficiently accurately for most practical purposes.

The difference between the two thermometer readings taken from the reading of the wet bulb, gives the dew-point very nearly, when the air is at any temperature between freezing and 80°. This simple rule will be found serviceable to horticulturists, since it will enable them to estimate the chilling effect of dew or hoarfrost on tender plants.

Use as an Indicator of Weather.—In our climate, the usual difference between the thermometer readings,—in the open air, shaded from the sun, reflected heat, and currents of air,—ranges from one to twelve degrees. In hot and dry climates, as India and Australia, the range out of doors has been found as much as 80°, occasionally.

When the moisture is frozen, the bulb should be wetted afresh, and the reading taken just before it again freezes; but the observation then is of little value, and for general purposes need not be taken, as the air is known to be dry in frosty weather.

The muslin or cotton rag should be washed once or twice a week by pouring water over the bulb; and it should be replaced by a fresh piece at least once a month. Accuracy depends very much upon keeping the wet bulb clean, and not *too* wet.

In connection with the barometer, this hygrometer is very useful, not only on land, but especially at sea, where other kinds of hygrometers cannot be practically used. A fall in the barometer is indicative of coming wind or rain: if the hygrometer shows increasing dampness by the difference of the readings becoming smaller,—rain may therefore be anticipated. On the contrary, if the hygrometer shows continuing or increasing dryness, a stronger wind is probable, without rain.

Domestic Uses.—Mason's hygrometer is useful in regulating the moisture of the air of apartments; a difference in the thermometer readings of from 5° to 8° being considered healthy. Many complaints require that the temperature and humidity of the air which the invalid breathes should be carefully regulated. Hence it is a valuable household instrument. In a room, it should be placed away from the fire as much as possible, but not exposed to draughts of air.

Figs. 80 and 81 show cheap arrangements of the instrument for domestic purposes. Other arrangements are given to the instrument to make it suitable for exhibiting the hygrometrical state of the air in hot-houses, conservatories, malting-houses, warehouses, manufactories, &c.

Fig. 82 shows the instrument arranged on brass tripod stand, with folding legs and metal cover, to render it portable.

Fig. 80.   Fig. 81.   Fig. 82.

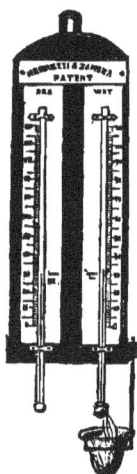 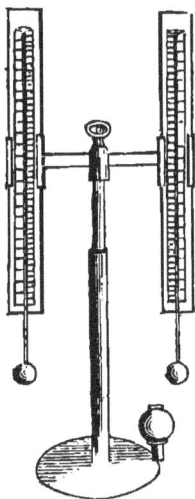 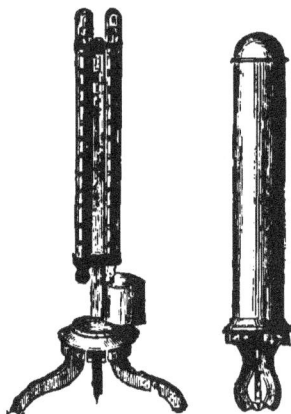

107. Self-Registering Hygrometer.—A maximum thermometer and a minimum thermometer, each fitted up as a wet-bulb thermometer, record the highest and lowest temperature of evaporation during the interval of observation. Negretti's mercurial maximum, and an alcohol minimum, answer best.

108. Causes of Dew.—"The aqueous vapour of our atmosphere is a powerful radiant; but it is diffused through air which usually exceeds its own mass more than one hundred times. Not only, then, its own heat, but the heat of the large quantity of air which surrounds it, must be discharged by the vapour, before it can sink to its point of condensation. The retardation of chilling due to this cause enables good solid radiators, at the earth's surface, to outstrip the vapour in their speed of refrigeration; and hence, upon these bodies, aqueous vapour may be condensed to liquid, or even congealed to hoar-frost, while at a few feet above the surface it still maintains its gaseous state."[*]   The amount of moisture so deposited will vary with different atmospheric conditions. If the sky be decidedly cloudy or misty, the heat radiated from the earth will be partly restored by counter-radiation from the visible vapour; the cooling of the earth's surface will, therefore, take place slowly, and little dew will be deposited. On the other hand, if the air contain transparent vapour, and the sky appear clear, the counter-radiation will be less, the earth will cool rapidly, and the deposit of dew will be copious; provided the night be comparatively calm, for, when the wind blows, the circulating air supplies heat to the radiating substances, and prevents any considerable chilling.

The dew which falls in tropical countries greatly exceeds in abundance what we experience in our climate; because the air is there, from the great heat, capable of

---

[*] Tyndall's *Heat considered as a Mode of Motion.*

sustaining a large amount of vapour in the transparent state, and the conditions most favourable for a maximum reduction of temperature by radiation are present. At those places, or upon those substances which cool the lowest and most readily, the dew falls most copiously.

109. Plan of Exposing Thermometers, &c.—Figure 83 is an illustration of a convenient slab for supporting thermometers in an exposed position attached to a stand (such as Glaisher's, described in Chapter XVI.) for ordinary scientific observations. It has a projecting ledge, *B*, to carry off rain from the instruments, the slab, *A*, being erected vertically. The hygrometer is placed at *E*, with the vase of water at *F*. An alcohol minimum thermometer is represented at *C*, in the position most favourable to its certain action; and at *D* is shown one of Negretti & Zambra's maximum thermometers, the position of which may be more nearly horizontal than there exhibited, although a slight depression of the bulb-end of the frame is desirable, but not necessary, as this thermometer can be used in any position.

Fig. 83.

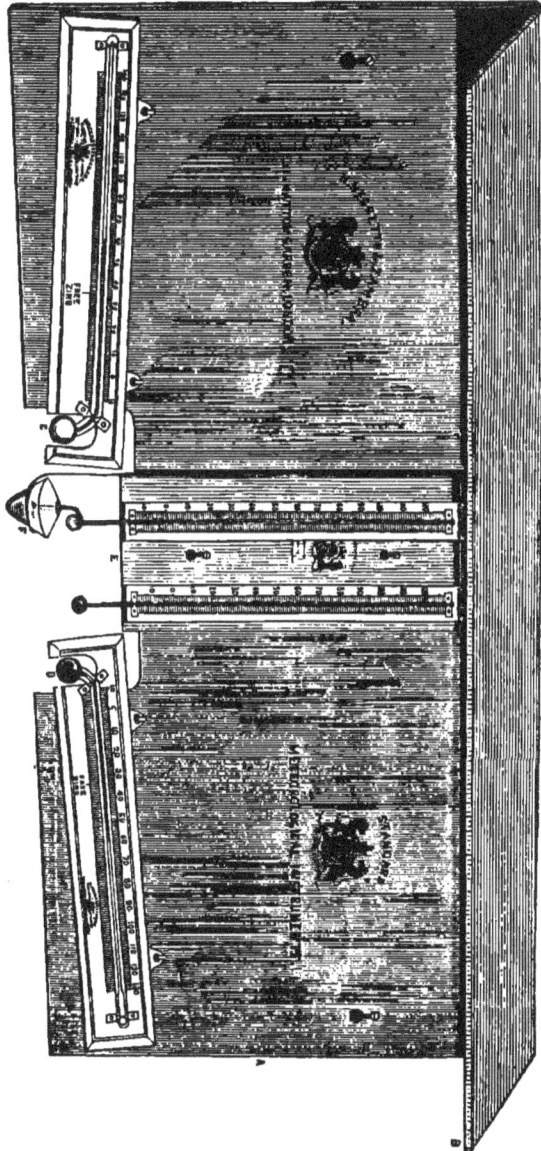

CHAPTER XII.

INSTRUMENTS USED FOR MEASURING THE RAINFALL.

The instruments in use for measuring the quantity of rain which falls on a given spot are of very simple construction. Perhaps the simplest is:—

110. Howard's Rain-Gauge.—It consists of a copper funnel, a stout glass or stone bottle, and a measuring glass. The bottle is to be placed upon the ground, with the funnel resting on its neck. A brass band or cylinder fixed upon the outer surface of the funnel envelops the neck of the bottle, and the pipe of the funnel extends nearly to the bottom of the bottle; so that loss by evaporation is avoided as much as possible. The receiving space of the funnel is formed by a brass ring, five inches in diameter, very accurately turned. The measuring vessel enables the observer to note the rainfall in inches, tenths, and hundredths of an inch.

111. Glaisher's Rain-Gauge.—The rain-gauge designed by Mr. Glaisher, the well-known meteorologist, and used by most observers of the present day, is arranged for the reception of the water which falls upon its receiving surface only, and for the prevention of loss by evaporation. The rain is first collected in a funnel, *B*, (fig. 84,) the receiving surface of which is turned in a lathe. The conical surface

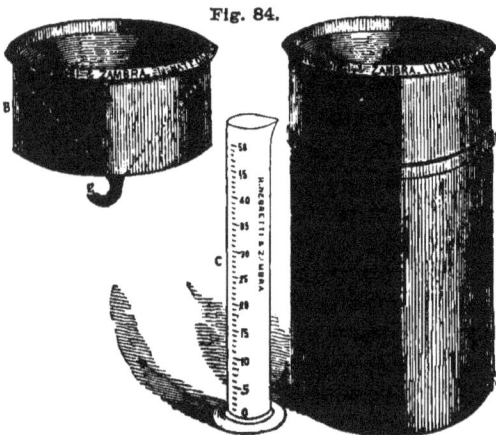

Fig. 84.

of the funnel slopes to the pipe, *E*, at an angle of 60° from the horizontal receiving surface. The tube, *E*, is of small aperture, and is bent up, in order to retain the last few drops of rain, so that the only opening for the escape of vapour may be closed as long as possible. The funnel, *B*, fits upon the cylinder, *A*, tightly in the groove, *D*. A copper can is placed inside the cylinder, *A*, to receive the rain from the funnel. Once or twice a day, or after a shower, this can should be taken out, and the water measured in the glass measure, *C*, which is graduated to hundredths of an inch, according to the calculated quantity of water, determined by the area of the receiving space. In use, this gauge should be partly sunk in the ground, so that the top may be about five inches above it. Thus situated, there will be little or no evaporation from it during any month of the year; and the readings need not be taken daily, although desirable.

112. Rain-Gauge with Float.—In this construction the graduated glass measure is dispensed with. The cylinder of the gauge is made less in diameter than the funnel, and a hollow, very flattened spheroid of copper forming a float, and carrying a vertical graduated boxwood scale which moves through the orifice of the funnel, is placed in it. As the rain accumulates the float rises, and the amount of rain in the gauge is read upon the scale from the top of the gauge, a bar, having a hole at the centre for the passage of the scale, being fixed diametrically across the receiving space of the funnel. The gauge is provided at the bottom with a brass cock, by which the water may be allowed to flow out of it whenever necessary.

This form of gauge is not very suitable for the measurement of small quantities; but is admirably adapted for localities where the rainfall is excessive.

113. Rain-Gauge with Side-Tube.—This instrument, as represented in fig. 85, is a cylindrical vessel, mounted on a base shaped as a frustum of a cone. This base may be filled with sand or gravel to make the instrument stable, so that when placed upon a lawn or in a garden it may have an ornamental appearance. The funnel for collecting the rain is larger in diameter than the cylinder. Parallel to the cylinder, and communicating with the lowest part of the interior and extending to its top, is a graduated glass tube, open at both ends. The rain collected will rise as high in this tube as in the cylinder, and its amount can therefore be read off without any trouble. The gauge is emptied by the brass tap at the bottom of the cylinder.

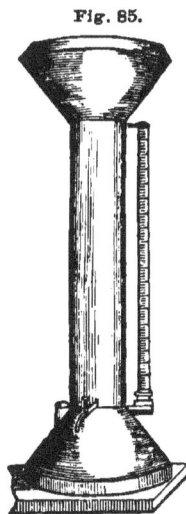
Fig. 85.

114. Admiral FitzRoy's Rain-Gauge.—A form of rain-gauge, very well adapted for expeditious observation at any time, has been designed by Admiral FitzRoy, and extensively employed by his observers. It is cylindrical in shape, with the funnel let into the top ; and the rainfall is collected in an inner and much smaller cylinder, so that a small fall is represented by a considerable depth of water in the gauge. The amount of rain which has fallen is ascertained by a dipping tube, similar in principle to the dipping syphon used by gaugers for taking out specimens of wines or spirits from casks by simply removing the bung. A short, vertical, tubular opening provided with a cap, which is attached to the instrument by a chain that it may not be lost, is formed in the funnel. The measuring tube, which has a small hole at each end, should be placed upright in the gauge; then the thumb should be pressed over the upper aperture, while the tube is lifted gently out, holding in the lower part a quantity of water representing the depth of the rain in the gauge, the upper edge of which is at the mark to be read off. The glass tube is graduated to inches and tenths; hundredths of an inch can be readily estimated by the eye. The marks are fixed by actual trial with a standard gauge, and are artificial, not true, inches.

115. Self-Registering Rain-Gauge.—The rain-gauge can be combined with clock-work and other mechanism so as to be self-recording of the amount of rain, the time, and duration of its fall. For the details of construction the reader is referred to the next chapter, where he will find the instrument described in connection with Osler's anemometer, as the "pluviometer." To observe and duly record the times of commencement and termination of rain is very desirable. Scarcely any observer can attempt to do this even approximately from personal observation. Hence the want of a cheap and simple self-recording rain-gauge is much felt, the present construction being too expensive for all but a few individuals.

In 1862, Mr. R. Strachan estimated the duration and amount of rain in London (Gray's Inn Road) as follows :—

Months.	Inches.	Days.	Hours.	Months.	Inches.	Days.	Hours.
January ...	1·86	19	88	July	2·27	17	68
February..	0·87	9	25	August	2·45	12	72
March.....	3·40	22	130	September.	1·70	12	55
April	2·34	14	80	October	3·23	21	94
May	3·04	16	90	November .	1·12	10	53
June	2·45	20	83	December .	1·44	17	66

" During the year 1862, the rainfall amounted to 25·67 inches. Rain fell on 179 days, that is, on nearly every other day. The hours of rain were estimated at 904 ; therefore, if the rain had fallen continuously, it would have lasted nearly 38 days and nights."[*] The value of similar estimates of the rainfall by numerous observers would be very great to meteorology.

116. The principle of measurement in all these gauges is the relation existing between the areas of the collecting and receiving surfaces; that is, between the area of the funnel into which the rain falls, and the area of the cylinder which receives it. In Howard's and Glaisher's gauges, this cylinder is virtually the measuring glass itself; in the others, above described, the measuring scales show the same depth of water as in the cylinder of the gauge.

The cylinder being of less diameter than the funnel, and receiving all the rain collected by the funnel, it follows that its contents will have an increased depth. Now equal cylindrical volumes, having different diameters, are to each other in

---

[*] Vide *Horological Journal*, Vol. V.

length inversely as the squares of the diameters. Hence, if the funnel be 9 inches and the cylinder 3 inches in diameter, a fall of 1 inch of rain will be represented in the gauge by 9 inches; for $3^2 : 9^2 :: 1 : x = 9$. In this case, therefore, a length of nine inches of the measuring glass, tube, or scale, would represent an inch of rainfall, and be divided into tenths and hundredths of the artificial inch.

117. Position for Rain-Gauge, &c.—Rain gauges should be placed on the ground, in any position exposed to a free fall of rain, snow, or hail, where neither walls, buildings, nor trees shelter or cause eddies of wind. They should be supported by a frame, or other means, to prevent them being blown down by the wind, but so that they can be readily emptied.

During snow or frost, the gauge must be watched, and its contents melted by placing it in a warm room, either when the amount is to be measured, or the funnel is filled up with snow. A tin vessel of equal area to the funnel may at such times be useful as a substitute.

Rain gauges are constructed of metal, usually copper, which, besides being readily workable, is little affected by atmospheric influences. If made of iron or zinc, they should be well japanned; if of copper, this is not so essential. The capacity of a gauge should be sufficient to contain at least the probable maximum fall of rain in a day at the locality. Those required for rainy districts must be of large size.

118. Causes of Rain.—When the invisible vapour which is diffused in the atmosphere becomes sufficiently cooled, it appears visible as mist or cloud, and a further reduction of temperature causes its precipitation as rain, hail, or snow. The cooling of the higher regions of the atmosphere is doubtless the chief cause of this condensation; but the property which aqueous vapour possesses of radiating heat may also contribute to the result. Moreover, the law which regulates the amount of vapour which air at any particular temperature can sustain in a transparent state, determines that when two bodies of air at different temperatures, saturated with vapour, intermix, some moisture must be rendered visible; and hence, it is not only possible, but highly probable, that rain may result from the conflict of different winds. Let us imagine two cubic yards of air, both saturated with moisture, but having the respective temperatures of 50 and 70 degrees, to come into contact. There will be a tendency to equalize the temperature to a mean, which is 60°; and during this process, some of the vapour will be condensed.

For in the air at 50° there is	... ...	110·7	grains of vapour*
and ,, 70 ,,	... ...	216·0	,,
Total amount of vapour ...	... ...	326·7	,,
But two cubic yards of air at 60° can only sustain		313·2	,,
Hence there will be deposited	... ...	13·5	,, of rain.

* *Hygrometrical Tables*, by J. Glaisher, Esq., F.R.S.

I

It may be conceded, therefore, that when a warm and moist current of air encounters a body of cold air which may not be extremely dry, the mixture is unable to retain the whole of the vapour in an invisible state; so that the excess becomes visible as mist or fog, and, when the temperature has become sufficiently lowered, rain. The British Isles are more or less enveloped in fog, or mist, at the commencement of easterly winds, which, with a sudden change of wind, is exhibited even in summer; while the south-westerly winds, warm, and arriving from the ocean, deposit large quantities of rain by the cooling effect of the land, colder by reason of its latitude. When rain occurs with a northerly wind, it is probably due to the deposition from an upper south-westerly current, often apparently proved by the movements of the upper clouds.

119. Laws of Rain-fall.—Tropical countries have a dry and a wet season during the year: *dry*, when the sun is at the opposite side of the equator; *wet*, when the sun is overhead. With reference to the British Isles, the statistics collected by Mr. G. J. Symons indicate that: 1st. The stations of least rain are inland, or on the east or south-east coasts; the stations of greatest rain are on the western coasts. 2nd. The rain-fall is very large in the vicinity of mountain chains or groups, unless the station happens to be some miles to the north-eastward.

It may be well to illustrate these remarks by quoting* the average fall at a few places, grouping them as—

Westerly.	Inches.	Central.	Inches.	Easterly.	Inches.
Bodmin	43	Enfield	23	Witham (Essex)	21
Bolton (Lancashire)	44	Epping	23	Patrington (Hull)	21
Coniston (Windermere)	71	Derby	24	Sunderland	17
Seathwaite	127	York	22	Inveresk (Edinburgh)	25
Torosay (I. of Mull)	75	Stirling	39	Pittenweem (Fife)	24
Killaloe (Limerick)	38	Perth	29	Dublin	22

Mr. Green, the celebrated aeronaut, has asserted from his experience, "that whenever a fall of rain happens, and the sky is entirely overcast, there will invariably be found to exist another stratum of cloud at a certain elevation above the former;" and the recent scientific balloon ascents by Mr. Glaisher have tended to confirm this theory. Mr. Glaisher says, "It would seem to be an established fact, that whenever rain is falling from an overcast sky, there is a second stratum above." "It would also seem that when the sky is overcast without rain, that there is no stratum of cloud above, but that the sun is shining on the upper surface. In every instance in which I have been up under these circumstances, I have found such to be the case, agreeing in this respect also with Mr. Green's observations."

---

* Vide *Report of the British Association*, 1862. It may be added, for the information of those who are about to commence observing, that Mr. Symons, of Camden Road Villas, London, is desirous of securing returns of rain-fall from as many stations as possible, in order to render more complete his annual reports to the British Association.

The amount of rain collected in a gauge placed near the surface of the earth is larger than in any gauge placed above it; and the higher the gauge is placed, the less water is collected. Mr. Glaisher contends that his balloon experiments corroborate this law.

120. Utility of Statistics of Rain-fall.—The utility of knowing the rain-fall of any locality is sufficiently obvious, and little need be said upon the subject. The rain-gauge should be in the hands of every gardener and farmer. In the management of out-door plants and crops, as well as in the construction of cisterns and tanks for the supply of water, a rain gauge is a valuable assistant. By its use, the gardener will be guided in judging how far the supply of moisture to the earth is needed ; and he will also see how beneficial is even a hasty shower to growing plants, when he considers that a fall of rain measuring the tenth of an inch in depth, corresponds to the deposit of about forty hogsheads per acre. The study of the rain-fall of a country is of considerable interest to agriculturists. The health and increase of domestic animals, the development of the productions of the land, as well as the daily labours of the farmer, are dependent upon the excess or deficiency of rain. " It must be a subject of great satisfaction and confidence to the husbandman to know at the beginning of a summer, by the certain evidence of meteorological results on record, that the season, in the ordinary course of things, may be expected to be a dry and warm one ; or to find, in a certain period of it, that the average quantity of rain to be expected for the month has fallen. On the other hand, when there is reason, from the same source of information, to expect much rain, the man who has courage to begin his operations under an unfavourable sky, but with good ground to conclude, from the state of his instruments and his collateral knowledge, that a fair interval is approaching, may often be profiting by his observations; while his cautious neighbour who waited ' for the weather to settle' may find that he has let the opportunity go by. This superiority, however, is attainable by a very moderate share of application to the subject ; and by the keeping of a plain diary of the barometer and rain-gauge, with the hygrometer and vane, under his daily notice."* The statistics of rain-fall are not only valuable and interesting in a meteorological point of view, and for agricultural purposes, but are also highly important in connection with sanitary arrangements for towns, and engineering operations. This is especially evident to the hydraulic engineer. As rain is an important source of water-supply to rivers, canals, and reservoirs, it is evident that a knowledge of the probable fall for any season or month, at a given place, as furnished by averages of the observations of former years, will be the data upon which the engineer will base his plans for providing for floods or droughts ; while the measurement of the actual quantity which has just fallen, as gathered from the indications of a series of gauges, will suggest to him the precautions to adopt either to economise or conduct away the in-pouring waters.

---

* Luke Howard's *Climate of London.*

" When a canal is conducted across an undulating country, its course is necessarily governed by the accidents of the ground, and it alternately rises and falls. In this case, rising by a succession of levels, it necessarily arrives at a certain highest level, which is called by engineers the *summit level*.   From this it again descends by a corresponding series of levels.   Now, it is evident that, supposing the locks to be all equal in magnitude, the ascent of a vessel will require the descent of as much water from the summit to the lowest level as would fill a single lock ; for this quantity of water must be discharged from each lock of the series when the vessel passes through it.

" The same may be said of the process by which the vessel descends along the series of locks on the other side of the summit.   It appears, therefore, that a supply of water must always be maintained on the summit level sufficient to fill a single lock twice for each vessel which crosses the summit.

" It happens, fortunately, that by the laws of natural evaporation, rain is precipitated in greater quantities on elevated summits than on the intermediate valleys, so that the moving power, in this case, accommodates itself to the exigencies of intercommunication."—*Dr. Lardner's* " *Handbook of Natural Philosophy.*"

121. New Form of Rain-Gauge.—Since the foregoing pages were in type, a modification of Howard's rain-gauge has been arranged by Mr. Symons, which is compact in design, convenient in use, and low in price.   It combines the advantages of most gauges ; having solidity, and facility of measurement.   The bottle is placed in a tin case, to the bottom of which are attached stout spikes, which, when forced into the earth, prevent its being upset either by wind or accident.   The bottle being transparent, and slits made in the case, the fall of rain is seen at a glance, or with a race-glass, from a window.   The funnel being attached to the cover of the case is thereby kept strictly horizontal, and the depth of rain can be accurately measured by lifting the bottle from its case and emptying it into a graduated glass jar.

The funnel of this gauge is a very deep cone, to prevent the rain drops outsplashing.   When properly placed, the receiving surface will be twelve inches above the ground, which experience has shown to be the most advantageous height.

CHAPTER XIII.

APPARATUS EMPLOYED FOR REGISTERING THE DIRECTION, PRESSURE, AND VELOCITY OF THE WIND.

122. The Vane.—The instrument by which the wind's direction is most gene-rally noted, is the vane, or weather-cock, and all that need be said of it here is that the points north, east, south and west, usually attached to it, should indicate the *true* and not the *magnetic* directions; and that care should be taken to prevent its setting fast. Very complicated instruments are required for ascertaining the pressure and velocity of the wind, and these are called *Anemometers.* The simplest is *Lind's.*

123. Lind's Anemometer, or Wind-Gauge (fig. 86), invented so late as 1775, for show-ing the pressure of the wind, consists of a glass syphon, the limbs parallel to each other, and each limb the same diameter. One end of the syphon is bent at right angles to the limb, so as to present a horizontal opening to the wind. A graduated scale, divided to inches and tenths, is attached to the syphon tube, reading either way from a zero point in the centre of the scale. The whole instru-ment is mounted on a spindle, surmounted by a vane, and is moved freely in any direction by the wind, always presenting the open end towards the quarter from which the wind blows. To use the instrument, it is simply filled up to the zero point with water, and then exposed to the wind; the difference in the level of the water gives the force of the wind in inches and tenths, by adding together the amount of depression in one limb, and elevation in the other, the *sum of the two* being the height of a column of water which the wind is capable of sustaining at that time.

Fig. 86.

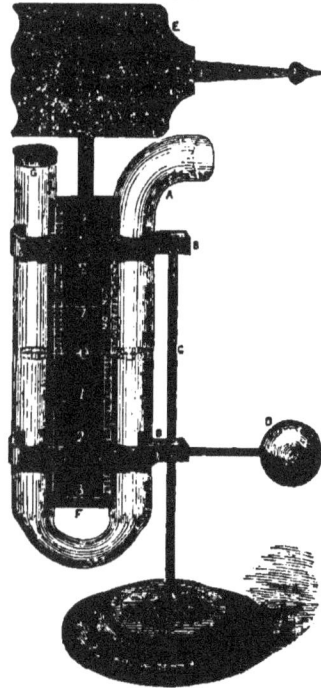

TABLE,

Showing the Force of Wind on a square foot, for different heights of the column of Water
in Lind's Wind-Gauge.

Inches.	Force in lbs.	Common designation of such Wind.
6	31·75	A Hurricane.
5	26·04	A violent Storm.
4	20·83	A great Storm.
3	15·62	A Storm.
2	10·42	A strong Wind.
1	5·21	A high Wind.
·5	2·60	A brisk Wind.
·1	·52	A fresh Breeze.
·05	·26	A gentle Breeze.
0.	0.	A Calm.

124. Modification of Lind's Gauge.—*Sir W. Snow Harris* has effected a
modification of Lind's anemometer, with a view of obtaining a hand instrument for use at sea more especially. At present the force of the wind is
estimated at sea by an arbitrary scale, suggested by Sir F. Beaufort, the late
hydrographer; 0 being calm, 12 the strongest hurricane, and the intermediate
numerals giving the varying strength of the wind. There has been a long-
felt want of instrumental means for obtaining this data at sea, if merely for the sake
of checking occasionally personal estimations, which may vary considerably among
different observers. Harris's wind gauge is intended to be held by hand, while facing
the wind, and keeping it in proper position by attending to a spirit-level attached.
When in position, and held firmly, the tube has to be opened to the wind by pressure of the thumb acting upon jointed levers, controlled by springs. The pressure
of the wind moves the enclosed liquid; and by withdrawing the thumb, the tube is
closed so as to keep the liquid in its position; the reading is then taken from its
scale, either in pounds on the square foot, miles per hour, or the ordinary desig-
nations of wind, as light, fresh, strong, &c.

125. Robinson's Anemometer.—*Dr. Robinson*, of Armagh, is the inventor
of a very successful anemometer, which determines the horizontal velocity of the
wind. It was first used in 1850, in the meteorological and tidal observations made
on the coast of Ireland under the direction of the Rev. Dr. Lloyd. No meteorological observatory should be without this valuable instrument, which is essential
in determining the average velocity of the wind of a locality as distinguished from

the most frequent wind of the same place. It is represented in fig. 87. Four hollow hemispherical cups, $A A$, are extended upon conjugate diameters, or arms, with their diametrical planes placed vertically, and facing the same way upon a vertical axis, B, which has at its lower extremity an endless screw, D. The axis is supported at C so as to turn with as little friction as possible. The endless screw is placed in gear with a train of wheels and pinions. Each wheel carries an index over a stationary dial in front; or the index is fixed, and the graduations are placed upon the wheels themselves.

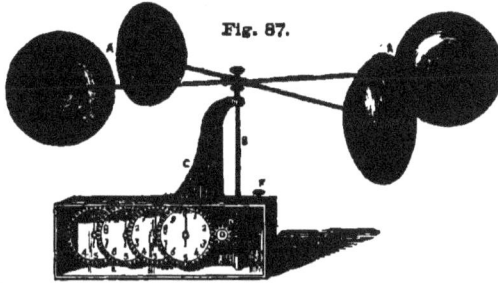

Fig. 87.

Dr. Robinson has proved, both by theory and experiment, that the centre of any one of the cups so mounted and set in motion by the wind, revolves with one-third of the wind's velocity. If, therefore, the diametrical distance between the centres of the cups be one foot, the circle described by the centres in one revolution is 3·1416 feet, and the velocity of the wind will be three times this, or 9·42 feet, which must be referred to time for the absolute rate. The instrument is sometimes made with the centres of the cups 1·12 feet apart, so that the circle described is $\frac{1}{1500}$ of a mile in circumference. Hence, to produce one revolution of the cups, the wind must travel three times as fast, or $\frac{1}{500}$ of a mile. Therefore, 500 revolutions will be produced by one mile of wind; so that the dials may be graduated to register the velocity in miles and tenths of miles. The simplest arrangement is with five dials, recording respectively 10, 100, 1,000, 10,000 and 100,000 revolutions.

Directions for using Robinson's Anemometer.—The dials read off in the same manner as the register of a gas meter, commencing with the dial farthest from the endless screw.

" The figures on the first dial indicate so many hundreds of thousands of revolutions; those on the second dial so many tens of thousands; those on the third, thousands; those on the fourth, hundreds; and those on the fifth so many tens.

" The instrument should be read every morning at 9 o'clock; and, usually, it will only be necessary to read the first three dials. The figures can be entered as they are read off. Should the index point *between* two figures, the less of the two is to be taken.

" For example, if the first dial points to 7, or between 7 and 8; while the second dial indicates 4 ; and the third, 5 ; the entry to be made is 745 (indicative of 745 *thousand* revolutions).

" Every time the index of the first dial is found to have passed zero (0), a cross or star is to be prefixed to the next (a lower) reading.

" To ascertain how many *thousands* of revolutions have been made during the month, it will simply be necessary to subtract the first reading from the last, and prefix to the three figures thus obtained a figure corresponding to the number of stars in the column. For every *thousand* revolutions there are two miles of wind : we have therefore only to multiply by 2 to find how many miles of wind have passed during the month.

" Two entries must be made for the last day of each month (the one being written under the other), so as to bring the readings down to 9 A.M. on the 1st of the following month. The same entry which ends one month, will therefore begin the next. This repetition of one entry is necessary, in order to prevent losing a day's wind.

" The accompanying example of the   687
readings of an Anemometer for 13 days   773
will illustrate the method of making   822
the entries, &c.   855

" In this instance, the first read-   900
ing (687) is less than the last (793).   953
When the first reading is greater than   990
he last, it will be necessary to borrow   *066
1,000 in making the subtractions,   197
and then deduct one from the num-   323
ber of stars. Thus, if the first read-   414
ing of the series on the margin had   597
been 887, the result would have been   712
906 instead of 1106.   793
                                 ————

1106 thousands of revolutions.

2

· 13 | 2212 miles of wind in period.
        ————

170 miles of wind per day, on an average.

" The foregoing directions are all which require to be regularly attended to. But it may be interesting at times to find the velocity of the wind during a period of a few minutes. This may be ascertained by observing the difference of two readings of all the dials, with an interval of some minutes between them, when a very brief calculation will suffice ; but perhaps the simplest method is the following :—

" Take two readings, with an interval of 12 minutes between them. The difference of these readings, divided by 10, is the velocity of the wind in miles per hour. Thus—if the reading of the five dials (from left to right) at noon is 15206, and at 12 minutes past 12 is 15348, the velocity of the wind is 14·2 miles per hour."—*Admiral FitzRoy, F.R.S.*

A lever and clutch are sometimes fitted to this anemometer, as in fig. 88, for throwing the train out of gear when not required to register. It may also be connected with clock-work so as to be self-recording, by causing the mechanism to impress a mark upon prepared paper moved by the apparatus, at certain intervals of time.

Fig. 88.

This anemometer should be fixed in an exposed situation, as high above ground as may be convenient for reading. It may be made very portable, by the arms which carry the cups being fitted to unscrew or to fold down. When fitted in gimbals, it can be used at sea with much advantage.

The pressure of the wind has been experimentally proved to vary as the square of the velocity; the relation being $V^2 = 200 \times P$. From this formula, therefore, the pressure can be calculated corresponding to the observed velocity.

126. Whewell's Anemometer.—This apparatus, the invention of the celebrated Dr. W. Whewell, registers the horizontal motion of the air with the direction. Its mechanism may be described in general terms, as follows:—

A horizontal brass plate is attached to a vertical spindle, which passes through the axis of a fixed cylinder, being supported by a bearing at the lower end, and working in a collar at the upper. A vane is attached, by which the plate is moved about according to the direction of the wind. A fly, having eight fans, each fixed at an angle of 45° with the axle, is placed upon the plate so that the axle is in the line of direction of the vane. An endless screw on the axle turns a vertical wheel having one hundred teeth, the axle to which has also an endless screw working into a horizontal wheel, having a like number of teeth, and which communicates motion to a vertical screw fifteen inches long. On this screw is placed a moveable nut, which carries a pencil. Round the cylinder is wrapped daily a paper divided for the points of the compass. The wind acting upon the vane will cause the plate to turn; and the screw which carries the pencil will travel with it, so that the pencil will mark upon the paper the direction of the wind. The fly will also be set in motion, and thereby the nut upon the screw will descend, so that the attached pencil will trace a vertical line upon the paper. When the fans on the axle are 2·3 inches from axis to end, and 1·9 inches wide, and the thread of the screw such that forty-five revolutions will cause the nut to descend two inches, 75·85 miles of wind will cause the pencil to descend through a vertical space of two inches; but the actual trace upon the paper will be longer in proportion to the magnitude of change of azimuth, or direction, of the wind.

127. Osler's Anemometer, and Pluviometer.—Mr. Follet Osler is the inventor of a self-recording apparatus which registers the direction and pressure of the wind, and the amount and duration of rain, upon the same sheet of paper. His apparatus has met with very much approbation, and has been erected in many observatories.

Fig. 89.

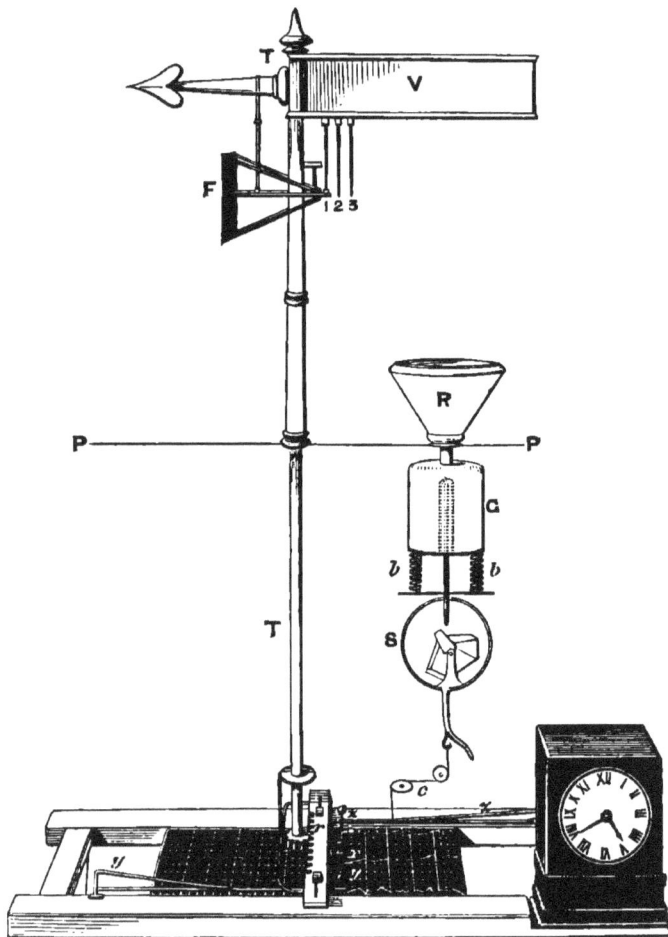

The mechanism may be modified in various ways, and the following is a description of the simplest and most recent arrangement.

The instrument, of which fig. 89 is a diagram rather than a picture, consists, first, of a vane, V, of a wedge-shape form, which is found to answer better than a flat vane; for the latter is always in a neutral line, and therefore is not sufficiently scnsitive. A wind-mill governor has been substituted for the vane to get the direction of the wind, with advantage. At the lower end of the tube, TT, is a small pinion, working in a rack, r, which moves backwards and forwards as the wind presses the vane. To this rack a pencil, x, is attached, which marks the direction of the wind on a properly ruled paper, placed horizontally beneath, and so adjusted as to progress at the rate of half an inch per hour, by means of a simple contrivance connecting it with a good clock. The paper is shown in the illustration upon the table of the instrument.

The pressure plate, F, for ascertaining the force of the wind, is one foot square, placed immediately beneath, and at right angles with the vane; it is supported by light bars, running horizontally on friction rollers, and communicating with flattened springs, 1, 2, 3, so that the plate, when affected by the pressure of the wind, acts upon them, and they transfer such action to a copper chain passing down the interior of the direction tube, and over a pulley at the bottom. A light copper wire connects this chain with the spring lever, $y\,y$, carrying a pencil which records the pressure upon the paper below. Mr. Osler much prefers a spring to any other means for ascertaining the force of the wind, because it is of the highest importance to have as little matter in motion as possible, otherwise the momentum acquired will cause the pressure plate to give very erroneous indications. The pressure plate is as light as is consistent with strength. It is kept before the wind by the vane, and is urged out by three or more springs, so that with light winds one only is compressed, and two, or more, according to the strength of the wind.

The *pluviometer* is placed on the right in the figure, $P\,P$ being the plane of the roof of the building. The rain funnel, R, exposes an area of about 200 square inches. The water collected in it is conveyed by a tube through the roof of the building into a glass vessel, G, so adjusted and graduated as to indicate a quarter of an inch of rain for every 200 square inches of surface, *i. e.* 50 cubic inches. G is supported by spiral springs, $b\,b$, which are compressed by the accumulating rain. A glass tube, open at both ends, is cemented into the bottom of G, and over it is placed a larger one closed at the top like a bell glass. The smaller tube thus forms the long leg of a syphon, and the larger tube acts as the short leg. The water, having risen to the level of the top of the inner tube, drops over into a little copper tilt, t, in the globe, S, beneath the reservoir. This tilt is divided into two equal partitions by a slip of copper, and placed upon an axis not exactly balanced, but so that one end or the other preponderates. The water then drops into the end of the tilt which happens to be uppermost, and when quite full it falls over, throwing the water into the globe, S, from which it flows away by the waste pipe. In this way an imperfect vacuum is produced in the globe, quite sufficient to produce a draught in the small tube of the syphon, or the long leg; and the whole contents of the reservoir, G, immediately run off, and the spiral springs, $b\,b$, elevate the reservoir to its original position. To produce this action, a quarter of an inch of rain must have fallen. The registration is easily understood. A spring lever, z, carrying a pencil, is attached by a cord, *c*, to *S*. This spring always keeps the cord tight, so that as the apparatus descends during the fall of rain, the spring advances the pencil more and more from the zero of the scale upon the paper beneath, until a quarter of an inch has fallen, when the pencil is drawn back to zero by the ascent of the reservoir.

The clock movement carries the registering paper forward by one of the wheels working into a rack attached to the frame.

The adjustment of the instrument should be carefully made at its first erection. The scale for pressure should be established experimentally, by applying weights of 2, 4, 6, &c., ℔s., to move the pressure plate.

The registration trace for twenty-four hours is readily understood. The direction is recorded on the centre part; the pressure on one side, and the rain on the other. Lines parallel to the length of the paper show no rain, steady wind, and constant pressure. On the rain trace, a line parallel to the width of the paper shows that the pencil had been drawn back to zero, a quarter of an inch of rain having fallen. The hour lines are in the direction of the width of the paper.

At the International Exhibition 1862, Messrs. Negretti and Zambra exhibited an improved Osler's anemometer, having combined with it Robinson's cups, so that the pressure and velocity appear on the same sheet, on which a line an inch in length is recorded at every ten miles; thus the complete instrument shows continuously the direction, pressure, and velocity of the wind.

128. Beckley's Anemometer.—Mr. R. Beckley, of the Kew Observatory, has devised a self-registering anemometer, which consists of three principal parts: Robinson's cups for the determination of velocity; a double fan, or wind-mill governor, for obtaining the direction; and a clock to move a cylinder, around which registration paper is wrapped. The paper records the time, velocity, and direction of the wind for twenty-four hours, when it must be replaced. It has a cast-iron tubular support, or pedestal to carry the external parts—the cups and the fans,—which must be erected upon the roof of the building upon which it is desired to mount the instrument.

The fans keep their axis at right angles to the wind; and with any change of direction they move, carrying with them an outer brass tube, which rests upon friction balls on the top of the pedestal, and is attached to a tubular shaft passing through the interior of the pedestal, and terminating with a mitre wheel. The mitre wheel, working with other cogged wheels, communicates the motion of the direction shaft to a cylinder carrying a pencil, to record the direction.

The shaft carrying the cups is supported upon friction balls, placed in a groove formed on the top of the direction shaft, and passing through the interior of that shaft, comes out below the mitre wheel, where it is terminated in an endless screw, or worm.

Upon the wind moving the cups, motion is given to the innermost shaft, thence to the worm-wheel, whence motion is given to a pencil which registers the velocity.

De la Rue's metallic paper is used in registration, it having the property of receiving a trace from a brass pencil. The pencils can, therefore, be made in the most convenient form. Mr. Beckley forms each pencil of a strip of brass wrapped round a cylinder, making a very thin threaded screw, so that the contact of the pencil cylinder and the clock cylinder is a mere point of the metallic thread. The pencil cylinders are placed side by side upon the cylinder turned by the clock, and require no spring or other appliance to keep them to their work, but always make contact with the registration paper by their own gravity. They therefore require no attention, and being as long as the trace which they make, they will last a long time.

The velocity pencil has only one turn on the cylinder, and its pitch is equal to a scale of fifty miles upon the paper. The direction pencil has likewise one turn on its cylinder, its pitch being equal to a scale of the cardinal points of the compass upon the paper.

The clock gives a uniform motion of half an inch per hour to the cylinder upon which the paper is fastened.

The registering mechanism of the instrument is very compact, requiring only a space of about 18 inches by 8 inches.

In the Report of the British Association for 1858, Mr. Beckley has given a detailed description of his anemometer, with drawings of all the parts.

129. Self-Registering Lind's Anemometer.—A Lind's wind-gauge, designed to register the maximum pressure, was exhibited at the International Exhibition 1862, by Mr. E. G. Wood. The bend of the syphon is contracted to obtain steadiness. On the leeward limb a hole is drilled corresponding in size with the contracted portion of the tube. The edge of the hole corresponds with the zero of the scale. On the pressure of the wind increasing, as much of the water as would have risen above the aperture flows away, and therefore the quantity left indicates the greatest pressure of the wind since the last setting of the instrument, which is done by filling it with water up to the zero point.

130. Anemometric Observations.—To illustrate the value of anemometric observations, we quote from a paper by Mr. Hartnup, on the results obtained from Osler's Anemometer, at the Liverpool Observatory. The six years' observations, ending 1857, gave for the yearly average of the winds: North-easterly, on 60 days, at 7·8 miles per hour; North-westerly, on 112 days, at 15·4 miles per hour; South-easterly, on 115 days, at 11·0 miles per hour; South-westerly, on 77 days, at 13·8 miles per hour; and one day calm. From the same observations, the average variation in the strength of the wind during the 24 hours is:—11 miles per hour, the minimum force, occurring at 1½ a.m.; until 6 a.m. it remains much the same, being then 11·3 miles per hour; at 10 a.m. it is 13·4 miles per hour; at 1½ p.m.

the wind is at its maximum strength, being 14·8 miles per hour; at 5 p.m. it is again 13·4 miles per hour, and at 9 p.m. 11·3 miles per hour. Hence it appears that the wind falls to its minimum force much more gradually than it rises to its maximum; that the decrease and increase are equal and contrary, so that the curve is symmetrical; and that generally the force of wind is less at night than during the day.

" There is evidence," says Admiral FitzRoy, " in Mr. Hartnup's very valuable anemometrical results, which seems to prove that to his observatory, in a valley, with buildings and hills to the north-eastward, the real polar current does not blow from N.E., but nearer S.E. By his reliable digest of winds experienced there, it appears that those most prevalent were from W.N.W. and S.S.E. But in England, generally, the *prevailing* winds are *believed* to be westerly, inclining to south-westerly, and north-easterly; while of all winds, the south-easterly is about the rarest.

" At Lord Wrottesley's observatory, in Staffordshire, about 520 feet above the sea, there appears to be considerably less strength of wind at any given time, when a gale is blowing *generally*, than occurs simultaneously at places along the sea-coast: whence the inference is, that undulations of the land's surface and hills, diminish the strength of wind materially by frictional resistance.

" All the synoptic charts hitherto advanced at the Board of Trade exhibit a marked diminution of force inland compared with that on the sea-coast. Indeed, the coast itself offers similar evidence, in its stunted, sloping trees, and comparative barrenness."*

---

* Vide *Third Number of Meteorological Papers*, issued by the Board of Trade.

CHAPTER XIV.

INSTRUMENTS FOR INVESTIGATING ATMOSPHERIC ELECTRICITY.

131. Atmospheric Electroscope.—The simplest instrument for ascertaining at any time the electric condition of the atmosphere is an electroscope composed of two equal pieces of gold leaf, suspended from a brass support, and insulated, as well as protected from the movement of the air, by a glass covering. Fig. 90 represents such an instrument. The cap of the brass support is fitted for the reception, in the vertical direction, of a metallic rod, not less than two or three feet in length. The top of the rod carries a clip. The instrument acts according to the law, that bodies similarly electrified repel each other; but when dissimilarly electrified, they attract each other. To make an observation, the instrument is placed in the open air, and a lighted piece of cigar fusee, or touch-paper, is fixed in the clip. The electricity of the air is collected by the substance undergoing combustion, and conducted by the rod to the gold leaf; and the pieces, being similarly electrified, separate more or less according to the amount of electricity present. The kind is determined by the effect of either an excited stick of sealing-wax or rod of glass upon the electrified gold leaf. A rod of glass, when rubbed briskly with a silk handkerchief or piece of woollen cloth, becomes positively electrified, or excited, as it is termed. A stick of sealing-wax, similarly treated, acquires the negative state. If, therefore, an excited glass rod be presented to the cap of the instrument, and it cause the pieces of gold leaf to diverge still further, the electric state of the air must be analogous to that of the glass, that is, *positive;* if they approach, it is *negative.* On the contrary, if a stick of sealing-wax be used, the pieces will be repelled more apart if they have acquired negative electricity from the air; and they will converge if they have a positive charge.

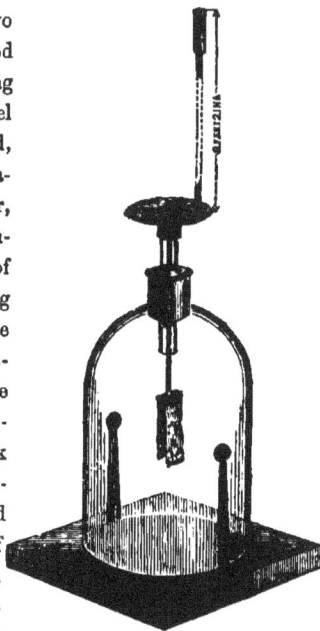

Fig. 90.

By means of this very simple instrument, meteorological observers can readily ascertain the electric condition of the lower air at any time.

NOTE.—A book containing strips of gold leaf is sent with the Electrometer to replace the gold leaves when torn or broken in use.

To mount fresh gold leaves, unscrew the brass plate to which is attached the rod supporting the leaves; then moisten with the breath the flat piece of brass, and press it gently down on one strip of gold, whilst the book is only partly opened; the second leaf is attached in the same manner.

132. Volta's Electrometer is similar to the instrument just described, except that instead of gold leaf two light pieces of straw, or two pith balls, are freely suspended from the conductor; the amount of the electric charge being estimated from the degrees of divergence, shown by a graduated arc.

133. Peltier's Electrometer is a much superior instrument in point of sensibility. A tall glass tube an inch or more in diameter, is connected to a glass receiver, mounted on a base fitted with levelling screws. At the top of the tube is formed a globe from four to five inches in diameter, which is thickly gilt on the exterior, so as to form a good conducting surface. A wire passes from the ball down the tube into the receiver, where it is bent up, and ends in a steel point over the centre of the base. A bent wire, carrying a small magnetic needle, is balanced on the steel point, so that the magnet, with the fine wire, arranges itself horizontally in the direction of the magnetic meridian. If any cloud or portion of air in the neighbourhood be in an electrical state, it will act by induction upon the gilt ball, and the needle will be deflected from its north and south direction.

A graduated circle indicates the number of degrees of the deflection, which will be greater or less according to the tension of the electricity. To ascertain whether the electricity is positive or negative, a stick of shellac or glass must be employed, as already described.

134. Bohnenberger's Electroscope may be fitted with a metallic conductor, and used with great advantage for observing atmospheric electricity. "The principal parts of the instrument, as improved by Becquérel, are the

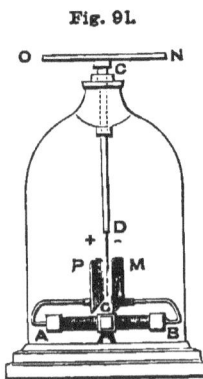

Fig. 91.

following:—A B, fig. 91, is a small dry galvanic pile of from 500 to 800 pairs, about a quarter of an inch in diameter; when the plates are pressed together, such a pile will be from 2 to 2½ inches in length. The wires, which are bent so as to stand above the pile, terminate in two plates, P and M, which are the poles of the pile. These plates, which are 2 inches by ¼ an inch, are parallel and opposite to each other. It is convenient for their opposite sides to be slightly convex, for them to be gilded or coated with platinum, and for them to run on the polar wires, by the latter being made to pass through a small hole in them. One of these plates will always be in a state of positive, and the other of negative, electricity; between them suspend the very fine gold leaf, $D G$, which is attached to the conductor, $C D$, of copper wire. If the leaf hang exactly between the two plates, it is equally attracted by each, and will therefore be in a state of repose. The apparatus should be protected by a bell-glass, fitting exactly, and having an opening at the top through which the copper wire, $C D$, passes; the wire, however, is insulated by its being contained in a glass tube, which is made to adhere to the bell-glass by means of a small portion of shellac or gum-lac. Screw on a metal ball or plate, to impart to it the electricity you wish to test, which will be conveyed by the copper wire to the gold leaf, and the latter will immediately move towards the plate which has the opposite polarity. This electroscope is, beyond doubt, one of the most delicate ever constructed, and is well adapted to show small quantities of positive and negative electricity.

" To ensure the susceptibility of electroscopes and electrometers placed under bell-glasses, precautions should be taken to render the air they contain as dry as possible, which may be effected by enclosing in a suitable vessel a little melted chloride of calcium beneath the glass."

The galvanic pile employed in this electroscope is that invented by Zamboni. " It differs from the common hydro-electric batteries principally in this, that the presence of the electromotive liquid is dispensed with, and that in its place is substituted some moist substance of low conducting power, generally paper. The electromotors in these piles are composed for the most part of Dutch gold (copper) and silver (zinc) paper pressed one on the other, with their paper sides together, out of which discs are cut with a diameter of from a quarter of an inch to an inch. More powerful pairs of plates may be obtained by using only the silver paper and smearing its paper side with a thin coat of honey, on which some finely pulverized peroxide of manganese has been sprinkled, and all the sides similarly coated are presented one way. Powerful pairs of plates may also be made by pasting pure gold leaf on the paper side of zinc-paper. These plates are then to be arranged, just as in the ordinary voltaic pile, one above the other, so that the similar metallic surfaces may all lie one way; press them tightly together; tie them with pretty stout silk threads, and press them into a glass tube of convenient size. The metal rims of the tubes, which must be well connected with the outermost pairs of plates, form the poles of the pile, the negative pole being in the extreme zinc surface, and the positive in the extreme copper or manganese surface.

" The electromotive energy called into action in these dry piles is less than that excited in the moist or hydro-electric piles, principally on account of the imperfect conduction of the paper. The accumulation of electricity at their poles also goes on less rapidly, and consequently the electrical tension continues for a long while unaltered; whereas, in all moist piles, even in the most constant of them, the tension is maintained, comparatively speaking, for but a short time, on account of

K

the chemical action and decomposition of the electromotive fluid—causes of disturbance which do not exist in the dry pile."*

135. Thomson's Electrometer.—Professor W. Thomson, of Glasgow, has devised an atmospheric electrometer, which is likely to become eminently successful, in the hands of skilful observers. It is mainly a torsion balance combined with a Leyden-jar. The index is an aluminium needle strung on a fine platinum wire, passing through its centre of gravity, and stretched firmly between two points. The needle and wire are carefully insulated from the greater part of the instrument, but are in metallic communication with two small plates fixed beside the two ends of the needle, and termed the repelling plates. A second pair of larger plates face the repelling plates, on the opposite side of the needle, but considerably farther from it. These plates are in connection with the inner coating of a Leyden-jar, and are termed the attracting plates. The whole instrument is enclosed in a metal cage, to protect the glass Leyden-jar and the delicate needle.

The Leyden-jar should be charged when the instrument is used. Its effect is two-fold: it increases greatly the sensibility of the instrument, and enables the observer to distinguish between positive and negative electrification.

The air inside the jar is kept dry by pumice-stone, slightly moistened with sulphuric acid; by which means very perfect insulation is maintained.

Electrodes, or terminals, are brought outside the instrument, by which the Leyden-jar can be charged, and the needle system connected with the body, the electric state of which is to be tested.

For the purpose of testing the electric state of the atmosphere, the instrument is provided with a conductor and support for a burning match, or, preferably, with an arrangement termed a water-dropping collector; by either of which means the electricity of the air is conveyed to the needle system.

The needle abuts upon the repelling plates when not influenced by electricity, in which position it is at zero. It can always be brought back to zero by a torsion-head, turning one end of the platinum wire, but insulated from it, and provided with a graduated circle, so that the magnitude of the arc, that the torsion-head is moved through to bring the needle to zero, measures the force tending to deflect it.

The action of the instrument is as follows:—The Leyden-jar is to be highly charged, say negatively; and the repelling plates are to be connected with the earth. The needle will then be deflected against a stop, under the combined influence of attraction from the Leyden-jar, or attracting plates, and repulsion from the repelling plates due to the positive charge induced on the needle and its plates by the Leyden-jar plates. The platinum wire must then be turned round by the torsion-head so as to bring back the needle to zero; and the number of degrees of torsion required will measure the force with which the needle is attracted. Next, let the needle

---

Elements of Physics, by C. F. Peschel.

plates be disconnected from the earth, and connected with the insulated body, the electric state of which is to be tested. In testing the atmosphere, the conductor and lighted match, or water-dropping apparatus, must be applied.

If the electricity of the body be positive, it will augment the positive charge in the needle plates, induced by the Leyden-jar plates ; and consequently the needle will be more deflected than by the action of the jar alone. If the electricity of the body be negative, it will tend to neutralize the positive charge; and the needle will be less deflected. Hence the kind of electricity present in the air becomes at once apparent, without the necessity of an experimental test. The platinum wire must then be turned till the needle is brought to zero, and the number of degrees observed ; which is a measure of the intensity of the electrification.

Any loss of charge from the Leyden-jar which may from time to time occur, reducing the sensibility inconveniently, may be made good by additions from a small electrophorus which accompanies the instrument.*

The instrument may be made self-recording by the aid of clockwork and photography. To effect this, a clock gives motion to a cylinder, upon which photographic paper is mounted. The needle of the electrometer is made to carry a small reflector ; and rays from a properly adjusted source of light are thrown by the reflector, through a small opening, upon the photographic paper. It is evident, that as the cylinder revolves, a trace will be left upon the paper, showing the magnitude of, and variations in, the deflection of the needle.

136. Fundamental Facts regarding Atmospheric Electricity.—The *general* electrical condition of the atmosphere is *positive* in relation to the surface of the earth and ocean, becoming more and more positive as the altitude increases. When the sky is overcast, and the clouds are moving in different directions, it is subject to great and sudden variations, changing rapidly from positive to negative, and the reverse. During fog, rain, hail, sleet, snow, and thunderstorms, the electrical state of the air undergoes many variations. The intensity of the electricity increases with hot weather following a series of wet days, or of wet weather coming after a continuance of dry days. The atmospheric electricity, in fact, seems to depend for its intensity and kind upon the direction and character of the prevailing wind, under ordinary circumstances. It has an annual and a diurnal variation. There is a greater diurnal change of tension in winter than in summer. By comparing observations from month to month, a gradual increase of tension is perceived from July to February, and a decrease from February to July. The intensity seems to vary with the temperature. The diurnal variation exhibits two periods of greatest and two of least intensity. In summer, the *maxima* occur about 10 a.m. and 10 p.m. ; the *minima* about 2 a.m. and noon. In winter, the *maxima* take place near 10 a.m. and 8 p.m. ; the *minima* near 4 a.m. and 4 p.m.

The researches of Saussure, Beccaria, Crosse, Quètèlet, Thompson, and

---

* This description is mod'fied from that in Report of the Jurors for Class XIII. International Exhibition, 1862.

FitzRoy have tended to show that during the prevalence of polar currents of air positive electricity is developed, and becomes more or less active according to the greater or less coldness and strength of wind; but with winds from the equatorial direction there is little evidence of sensitive electricity, and when observable, it is of the negative kind. Storms and gales of wind are generally attended, in places, with lightning and thunder; and as the former are very often attributed to the conflict of polar and equatorial winds, the difference of the electric tension of these winds may account for the latter phenomena. It is not our intention to enter upon the general consideration of thunderstorms; the facts which we have given may be of service to the young observer; and finally, as it is interesting to be able to judge of the locality of a thunderstorm, the following simple rule will be of service, and sufficiently accurate:—Note by a second's watch the number of seconds which elapse from the sight of the lightning to the commencement of the thunder; divide them by five, and the quotient will be the distance in miles. Thus, if thunder is heard ten seconds after the lightning was seen, the distance from the seat of the storm will be about two miles. The interval between the flash and the roll has seldom been observed greater than seventy-two seconds.

137. Lightning Conductors.—"The line of danger, whether from the burning or lifting power of lightning, is the line of strong and obstructed currents of air, of the greatest aerial friction."* Trees, church spires, wind-mills and other tall structures, obstruct the aerial currents, and hence their exposure to danger. The highest objects of the landscape, especially those that are nearest the thunder cloud, will receive the lightning stroke. The more elevated the object, the more likely is it to be struck. Of two or more objects, equally tall and near, the lightning is invariably found to select the best conductor of electricity, and even to make a circuitous path to get to it. Hence the application and evident advantage of metallic rods, called *lightning conductors*, attached to buildings and ships. A lightning conductor should be pointed at top, and extend some feet above the highest part of the edifice, or mast. It should be made of copper, which is a better conducting medium than iron, and more durable, being less corrosive. It must be unbroken throughout its length, and extend to the bottom of the building, and even some distance into the ground, so as to conduct the electricity into a well or moist soil. If it be connected with the lead and iron work in the structure of the house, it will be all the better, as affording a larger surface, and a readier means of exit for the fluid. In a ship, the lower end of the conductor should be led into communication with the hull, if of iron, and with the copper sheathing, if a wooden vessel; so that, spread over a large surface, it may escape more readily to the water.

138. Precautions against Lightning.—Experience seems to warrant the assumption that any building or ship, fitted with a substantial lightning conductor,

* *All the Year Round*, No. 224.

is safe from danger during a thunderstorm. Should a house or vessel be undefended by a conductor, it may be advisable to adopt a few precautions against danger. In a house, the fire-place should be avoided, because the lightning may enter by the chimney, its sooty lining being a good conductor. "Through chimneys, lightning has a way into most houses; and therefore, it is wise, by opening doors or windows, to give it a way out. Wherever the aerial current is fiercest, there the danger is greatest; and if we kept out of the way of currents or draughts, we keep out of the way of the lightning."* Lightning evinces as it were a preference for metallic substances, and will fly from place to place, even out of the direct line of its passage to the earth, to enter such bodies. It is therefore well to avoid, as much as possible, gildings, silvered mirrors, and articles of metal. The best place is perhaps the middle of the room, unless a draught passes, or a metallic lamp or chandelier should be hanging from the ceiling. The neighbourhood of bad conductors, such as glass windows, not being open, and on a thick bed of mattrasses, are safe places. The quality of trees as lightning conductors is considered to depend upon their height and moisture, those which are taller and relatively more humid being struck in preference to their fellows; therefore, it is unwise to seek shelter under tall and wet trees during a thunderstorm. In the absence of any other shelter, it would be better to lie down on the ground.

---

* *All the Year Round*, No. 224.

CHAPTER XV.

OZONE AND ITS INDICATORS.

139. Nature of Ozone.—During the action of a powerful electric machine, and in the decomposition of water by the voltaic battery, a peculiar odour is perceptible, which is considered to arise from the generation of a substance to which the term *ozone* has been given, on account of its having been first detected by smell, which, for a long time after its discovery, was its only known characteristic. A similar odour is evolved by the influence of phosphorus on moist air, and in other cases of slow combustion. It is also traceable, by the smell, in air,—where a flash of lightning has passed immediately before. Afterwards it was established that the same element possessed an oxidising property. It was found to be liberated at the oxygen electrode when water was decomposed by an electric current; and has been regarded by some chemists as what is termed an *allotropic* form of oxygen, while others speak of it as oxygen in the *nascent* state, and some even regard it as intimately related to chlorine. So various are the existing notions of the nature of this obscure agent.

Its oxidising property affords a ready means for its detection, even when the sense of smell completely fails. The methods of noting the presence and measuring the amount of ozone present in the air, are very simple; being the free exposure to the air, defended from rain and the direct rays of the sun, of prepared test-papers. There are two kinds of test-papers. One kind was invented by Dr. Schonbein, the original discoverer of ozone; and the other, which is more generally approved, by Dr. Moffat.

140. Schonbein's Ozonometer consists of strips of paper, previously saturated with a solution of starch and iodide of potassium, and dried. The papers are suspended in a box, or otherwise properly exposed to the air, for a given interval, as twenty-four hours. The presence of ozone is shown by the test-paper acquiring a purple tint when momentarily immersed in water. The amount is estimated by the depth of the tint, according to a scale of ten tints furnished for the purpose, which are distinguished by numbers from 1 to 10. The ozone decomposes the compound which iodine forms with hydrogen, and, it is presumed, combines as oxygen with hydrogen, while the iodine unites with the starch, giving the blue colour when moist.

141. Dr. Moffat's Ozonometer consists of papers prepared in a somewhat similar manner to Schonbein's; but they do not require immersion in water. The presence of ozone is shown by a brown tint, and the amount by the depth of tint according to a scale of ten tints, which is furnished with each box of the papers.

Moffat's have the advantage of preserving their tint for years, if kept in the dark, or between the leaves of a book; and are simpler to use.

142. Sir James Clark's Ozone Cage (fig. 92), consists of two cylinders of very fine wire gauze, one fitting into the other; the wire gauze being of such a fineness as to permit the free ingress of air, at the same time that it shuts out all light that would act injuriously on the test-paper, which is suspended by a clip or hook attached to the upper part of the inner cylinder.

Fig. 92.

143. Distribution and Effects of Ozone.—Mr. Glaisher has found that "the amount of ozone at stations of low elevation is small; at stations of high elevation, it is almost always present; and at other and intermediate stations, it is generally so. The presence and amount of ozone would seem to be graduated by the elevation, and to increase from the lowest to the highest ground. The amount of ozone is less in towns than in the open country at the same elevation; and less at inland than at sea-side stations." It seems to abound most with winds from the sea, and to be most prevalent where the air is considered the purest and most salubrious. This may seem, says Admiral FitzRoy, in *The Weather Book*, to point to a connection between ozone and chlorine gas, which is in and over sea-water, and which must be brought by any wind that blows from the sea. It prevails more over the ocean and near it than over land, especially land remote from the sea; and, says the Admiral, it affects the gastric juice, improves digestion, and has a tanning effect. Dr. Daubeny, in his *Lectures on Climate*, writes: "Its presence must have a sensible influence upon the purity of the air, by removing from it fœtid and injurious organic effluvia. It is also quite possible that ozone may play an important part in regulating the functions of the vegetable kingdom likewise; and although it would be premature at present to speculate upon its specific office, yet, for this reason alone, it may be well to note the fact of its frequency, in conjunction with the different phases which vegetation assumes, persuaded that no principle can be generally diffused throughout nature, as appears to be the case with this, without having some important and appropriate use assigned for it to fulfil."

144. Registering Ozonometer.—Dr. E. Lancaster has contrived an ozonometer, the object of which is to secure the constant registration of ozone, so that the varying quantities present in the atmosphere may be detected and registered. For this purpose, an inch of ozone paper passes in each hour, by clock-work, beneath an opening in the cover of the instrument.

CHAPTER XVI.

INSTRUMENTS NOT DESCRIBED IN THE PRECEDING CHAPTERS.

145. Chemical Weather Glass.—This curious instrument appears to have been invented more than a hundred years ago, but the original maker is not known. It is simply a glass vial about ten inches long and three quarters of an inch in diameter, which is nearly filled, and hermetically sealed, with the following mixture :—Two drachms of camphor, half a drachm of nitrate of potassium, half a drachm of chlorate of ammonium, dissolved in about two fluid ounces of absolute alcohol mixed with two ounces of distilled water. All the ingredients should be as pure as possible, and each vial filled separately. When the instruments are made in numbers and filled from a common mixture, some get more than the due proportion of the solid ingredients, and consequently such glasses do not exhibit that uniformity of appearance and changes, that undoubtedly should accompany similar influencing circumstances. It is in consequence of a want of precision and fixed principle of manufacture, that these interesting instruments are not properly appreciated, and more generally used.

The glass should be kept quite undisturbed, exposed to the north, and shaded from the sun. Camphor is soluble in alcohol, but not in water, while both water and alcohol have different solvent powers, according to the temperature; hence, the solid ingredients being in excess for certain conditions of solution, depending upon temperature chiefly, and perhaps electricity and the action of light also, appear as crystals and disappear with the various changes that occur in the weather.

The various appearances thus presented in the menstruum have been inferred to prognosticate atmospheric changes. The following rules have been deduced from careful study of the glass and weather :—

1. During cold weather, beautiful fern-like or feathery crystallization is developed at the top, and sometimes even throughout the liquid. This is the normal state of the glass during winter. The crystallization increases with the coldness ; and if the structure grows downward, the cold will continue.

2. During warm and serene weather, the crystals dissolve, the upper and greater part of the liquid becoming perfectly clear. This is the normal state of the glass during summer. The less amount of crystallization, that is, the greater the clear portion of the liquid (for there is always some of the composition visible at the bottom), the greater the probability of continued fine dry weather.

3. When the upper portion is clear, and flakes of the composition rise to the top and aggregate, it is a sign of increasing wind and stormy weather.

4. In cold weather, if the top of the liquid becomes thick and cloudy, it denotes approaching rain.

5. In warm weather, if small crystals rise in the liquid, which still maintains its clearness, rain may be expected.

6. Sharpness in the points and features of the fern-like structure of the crystals, is a sign of fine weather; but when they begin to break up, and are badly defined, unsettled weather may be expected.

Admiral FitzRoy, in *The Weather Book*, writes of this instrument as follows:—
" Since 1825, we have generally had some of these glasses, as curiosities rather than otherwise; for nothing certain could be made of their variations until lately, when it was fairly demonstrated that if fixed undisturbed in free air, not exposed to radiation, fire, or sun, but in the ordinary light of a well-ventilated room, or, *preferably*, in the outer air, the chemical mixture in a so-called storm-glass varies in character with the *direction* of the wind—not its force, *specially* (though it *may* so vary in *appearance*, only from another cause, *electrical tension*).

" As the atmospheric current veers toward, comes from, or is only *approaching* from the polar direction, this chemical mixture—if closely, even microscopically watched—is found to grow like *fir*, *yew*, fern leaves, or hoar-frost—or like crystallizations.

" As the wind, or great body of air, tends more from the *opposite* quarter, the lines or spikes—all regular, hard, or crisp features—gradually diminish, till they vanish.

" Before, and in a continued southerly wind, the mixture sinks slowly downward in the vial, till it becomes shapeless, like melting white sugar.

" Before, or during the continuance of a northerly wind (polar current), the crystallizations are beautiful (if the mixture is correct, the glass a *fixture*, and duly *placed*); but the least motion of the liquid disturbs them.

" When the main currents meet, and turn *toward the west*, making *easterly* winds, stars are more or less numerous, and the liquid dull, or less clear. When, and while they *combine by the west*, making westerly winds, the liquid is clear, and the crystallization well-defined, without loose stars.

" While *any hard* or *crisp* features are visible below, above, or at the top of the liquid (where they form for polar winds), there is *plus* electricity in the air; a *mixture* of polar current co-existing *in that locality* with the opposite, or southerly.

" When nothing but soft, melting, sugary substance is seen, the atmospheric current (feeble or strong as it may be) is southerly with *minus* electricity, unmixed with, and *uninfluenced* by, the contrary wind.

" Repeated trials with a delicate galvanometer, applied to measure electric tension in the air, have proved these facts, which are now found useful for aiding, with the barometer and thermometer, in forecasting weather.

" Temperature affects the mixture much, but not solely; as many comparisons of winter with summer changes of temperature have fully proved.

"A confused appearance of the mixture, with flaky spots, or stars, in motion, and less clearness of the liquid, indicates south-easterly wind, probably strong to a gale.

"Clearness of the liquid, with more or less perfect crystallizations, accompanies a combination, or a contest, of the main currents, by the *west*, and very remarkable these differences are,—the results of these air currents acting on each other *from* eastward, or from an entirely opposite direction, the *west*.

"The glass should be wiped clean now and then,—and once or twice a year the mixture should be disturbed, by inverting and gently shaking the glass vial."

146. Leslie's Differential Thermometer.—A glass tube having a large bulb at each extremity, and bent twice at right angles, as represented in figure 93, con-

Fig. 93.

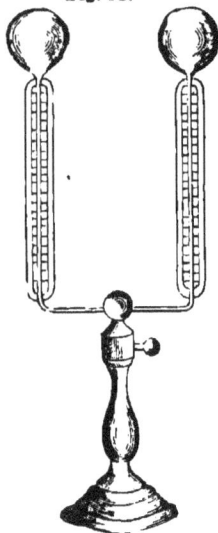

taining strong sulphuric acid tinged with carmine, and supported at the centre by a wooden stand, constitutes the differential thermometer as invented by Professor Leslie. The instrument is designed to exhibit and measure small differences of temperature. Each leg of the instrument is usually from three to six inches long, and the balls are about four inches apart. The calibre of the legs is about $\frac{1}{50}$ inch, not more; the other part of the tube may be wider. The tube is filled with the liquid, the bulbs contain air. When both bulbs are heated alike, each scale indicates zero. The scale is divided so that the space between the freezing and the boiling-points of water is equal to 1,000 parts. When one bulb is heated more than the other, the difference of temperature is delicately shown by the descent of the coloured fluid from the heated ball. It is uninfluenced by changes in the temperature of the atmosphere; hence it is admirably adapted for experiments of radiant heat. The theory of the instrument is that gases expand equally for uniform increments of heat.

147. Rumford's Differential Thermometer differs from that just described in simply containing only a small bubble of liquid, which lies in the centre of the tube, when both bulbs are similarly influenced. The bulbs and other parts of the tube contain air. When one bulb is more heated than the other, the bubble moves towards the one less heated; and the scale attached to the horizontal part of the tube affords a measurement of the difference of temperature.

148. Glaisher's Thermometer Stand,—The thermometer stand consists of a horizontal board as a base, of a vertical board projecting upwards from one edge of the horizontal one, and of two parallel inclined boards, separated from each other by blocks of three inches in thickness, connected at the top with the vertical, and at the bottom with the horizontal board, and the air passes freely about and between them all. To the top of the inclined boards is connected a small projecting roof to prevent the rain falling on the bulbs of the instrument, which are carried on the face of the vertical board, with their bulbs projecting below it, so that the air plays freely on the bulbs from all sides. The whole frame revolves on an upright post firmly fixed to the ground, as shown in the engraving, fig. 94; and in use, the inclined side is always turned towards the sun.

Fig. 94.

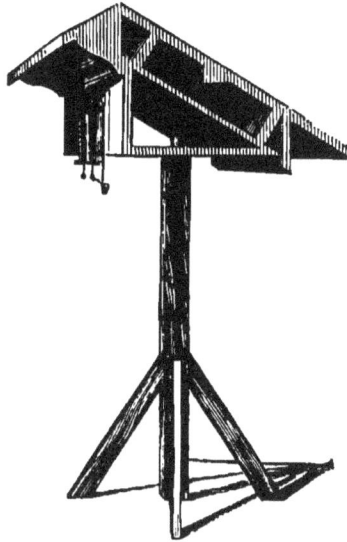

149. Thermometer Screen, for use at Sea.—This screen, or shade, was de-signed by Admiral FitzRoy, and has been in use for several years on board H.M. vessels and many merchant-ships. It is about twenty-four inches long by twelve wide and eight deep; having lattice-work sides, door, and bottom; with perforation also at top, so contrived that the air has free access to the interior, while the direct rays of the sun, rain, and sea spray are effectually excluded from the thermometers mounted inside. There is ample space for two thermometers placed side by side on brackets, at least three inches from each other or any part of the exterior of the screen. One thermo-meter should be fitted up as a " wet bulb" (see p. 105). A small vessel of water can easily be fixed inside the screen so as to retain its position and contents under the usual motions of the ship ; and by means of a piece of cotton-wick, or muslin rag tied round the bulb of the thermometer and trailing into the cup of water, keep the bulb constantly moist.

Self-registering thermometers should be protected by a similar screen. It has been found that thermometric observations made at sea are not valuable for scien-tific purposes unless the instru-ments have been duly protected by such a screen.

Fig. 95.

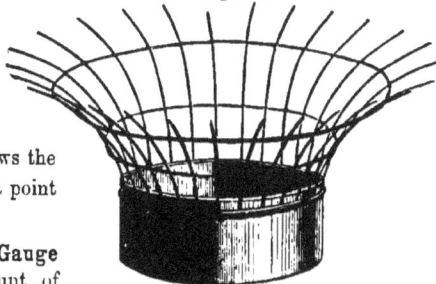

150. Anemoscope, or Port-able Wind Vane for travellers, with compass, bar needle, &c., shows the direct course of the wind to half a point of the compass.

151. Evaporating Dish, or Gauge (fig. 95), for showing the amount of evaporation from the earth's surface. This gauge consists of a brass vessel, the area or evaporating surface of which is accurately determined; and also a glass cylindrical measure, graduated into inches, tenths, and hundredths of inches. In use, the evaporating gauge is nearly filled with water, the quantity having been previously measured by means of the glass cylinder; it is then placed out of doors, freely exposed to the action of the atmosphere; after exposure, the water is again measured, and the difference between the first and second measurement shows the amount of evaporation that has taken place. If rain has fallen during the exposure of the gauge, the quantity collected by it must be deducted from the measured quantity; the amount is shown by the quantity of rain collected in the rain gauge. The wire cage round the gauge is to prevent animals, birds, &c., from drinking the water.

152. Dr. Babington's Atmidometer, or instrument for measuring the evaporation from water, *ice or snow*, consists of an oblong hollow bulb of glass or copper, beneath which and communicating with it by a contracted neck is a second globular bulb, duly weighted with mercury or shot. The upper bulb is surmounted by a small glass or metal stem, having a scale graduated to grains and half-grains; on the top of which is fixed horizontally a shallow metal pan. The bulbs are immersed in a vessel of water having a circular hole in the cover through which the stem rises. Distilled water is then gradually poured into the pan above, until the zero of the stem sinks to a level with the cover of the vessel. Thus adjusted, as the water in the pan evaporates, the stem ascends, and the amount of evaporation is indicated in grains. This instrument affords a means of measuring evaporation from *ice or snow*. An adjustment for temperature is necessary.

153. Cloud Reflector.—At the International Exhibition 1862, Mr. J. T. Goddard exhibited a cloud mirror, for ascertaining the direction in which the clouds are moving.

The mirror is laid on a horizontal support near a window, and fastened so that the point marked north may coincide with the south point of the horizon,—the several points will consequently be reversed. The edge of a conspicuous cloud is brought to the centre of the mirror, and the observer keeps perfectly still until it passes off at the margin, where the true point of the horizon *from which* the clouds are coming can be read off.

154. Sunshine Recorder.—Mr. Goddard also exhibited an instrument which he calls by this name. It works by letting the sun's rays pass through a narrow slit, and fall on photographic paper wound round a barrel moved by clock-work; the paper being changed daily, and the photographic impression developed and fixed in the usual manner.*

---

* Vide *Jurors' Reports.*

155. SET OF PORTABLE INSTRUMENTS.

In a small box, 8 in. by 8 in. by 4 in., a complete set of meteorological instruments have been packed. The lid of the box, by an ingenious arrangement, is made to take off and hang up; on it are permanently fixed for observation, a maximum and minimum, and a pair of dry and wet bulb thermometers. The interior of the box contains a maximum thermometer in vacuo for solar radiation, and a minimum for terrestrial purposes, one of Negretti and Zambra's small pocket aneroid barometers, pedometer for measuring distances, pocket compass, clinometer, and lastly a rain gauge. This latter instrument consists of an accurately turned brass ring having an india rubber body fastened to it to receive the rain, which is measured off by a small graduated glass, also contained in the box. Gentlemen travelling will find this compact observatory all that can be desired for meteorological observations.

156. IMPLEMENTS.

The practical meteorologist will find the following articles very useful, if not necessary. They scarcely require description; an enumeration will therefore suffice:—
Weather Diagrams, or prepared printed and ruled forms, whereon to exhibit graphically the readings of the various instruments to render their indications useful in fore-telling weather, &c. ;—*Meteorological Registers*, or Record Books, for recording all observations, and the deductions;—*Cloud Pictures*, by which the clouds can be readily referred to their particular classification, very necessary to the inexperienced and learners;—Cyclone Glasses, or Horns, outline Maps with Wind-markers, are also useful, especially in forecasting weather.

Fig. 96.

157. HYDROMETER.

Fig. 97.

A simple kind of hydrometer is very much used at sea, as "a sea-water test ;" and as the observations are usually recorded in a meteorological register or the ship's log-book, it may not be altogether out of place to give a description of it here.

It is constructed of glass. If made of brass, the corrosive action of salt-water soon renders the instrument erroneous in its indications. The shapes usually given to the instruments are shown in figs. 96 and 97. A globular bulb is blown, and partly filled with mercury or small shot, to make the instrument float steadily in a vertical position. From the neck of the bulb the glass is expanded into an oval or a cylindrical shape, to give the instrument sufficient volume for flotation; finally, it is tapered off to a narrow upright stem which encloses an ivory scale, and is closed at the top. The divisions on the scale read downward, so as to measure the length of the stem which stands above the surface of any liquid in which the hydrometer is floated. The denser the fluid, the higher will the instrument rise; the rarer, the lower it will sink.

The indications depend upon the hydrostatic principle, that floating bodies displace a quantity of the fluid which sustains them equal to their own weight. According, therefore, as the specific gravities of fluids differ from each other, so will vary the quantities of the fluids displaced by the same body when floated successively in each.

The specific gravity of distilled water, at the temperature of 62° F, being taken as unity, the depth to which the instrument sinks when gently immersed in such water is the zero of the scale. The graduations extend from 0 to 40; the latter being the mark which will be level with the surface when the instrument is placed in water, the specific gravity of which is 1·040. In recording observations, the last two figures only—being the figures on the scale—are written down. Sea-water usually ranges from 1·020 to 1·086.

A small tin, copper, or glass cylinder is useful for containing the water to be tested. It should be wider than the hydrometer, and always filled to the brim. If fitted to a stand, which is supported by gimbals, it will be very convenient. Water in a bucket, basin, or other wide vessel, acquires motion at sea, and the eye cannot be brought low enough (on account of the edges) to read off the scale accurately.

Errors of observation may occur with the hydrometer, if it be put into water without being clean, or without being carefully wiped. The instrument is extremely accurate if correctly used. It should be kept free from contact with the sides of the vessel; and all dust, smears, or greasiness, should be scrupulously avoided, by carefully wiping it with a clean cloth before and after use.

Whenever the temperature of the water tested differs from 62°, a correction to the reading is necessary, for the expansion or contraction of the glass, as well as the water itself, in order to reduce all observations to one generally adopted standard.

Negretti and Zambra's hydrometer, with thermometer in the stem, shows the density and temperature in one instrument.

For the following Tables we are indebted to the kindness of Admiral FitzRoy :—
TABLE for reducing observations made with a BRASS HYDROMETER, assuming the linear expansion of brass to be 0·000009555 for 1° F. The correction is additive for all temperatures above 62°, and subtractive for temperatures below 62°.

$t°$	Correction.	$t°$	Correction.	$t°$	Correction.	$t°$	Correction.
32	—0·0014	48	—0·0010	64	+0·0002	80	+0·0020
33	·0014	49	·0009	65	·0003	81	·0021
34	·0014	50	·0009	66	·0004	82	·0023
35	·0014	51	—0·0008	67	·0005	83	·0024
36	·0014	52	·0008	68	+0·0006	84	·0026
37	·0014	53	·0007	69	·0007	85	+0·0027
38	—0·0014	54	·0006	70	·0008	86	·0029
39	·0013	55	.0006	71	·0009	87	·0030
40	·0013	56	—0·0005	72	·0010	88	·0032
41	·0013	57	·0004	73	·0011	89	·0033
42	·0013	58	·0003	74	+0·0013	90	+0·0035
43	·0012	59	·0003	75	·0014	91	·0036
44	—0·0012	60	·0002	76	·0015	92	·0038
45	·0011	61	—0·0001	77	·0016	93	·0040
46	·0011	62	0.0000	78	·0018	94	·0041
47	—0·0010	63	+0·0001	79	+0·0019	95	+0·0043

TABLE for reducing observations made with a GLASS HYDROMETER, assuming the linear expansion of glass to be 0·00000463 for 1° F. The correction is additive for temperatures above 62°, and subtractive for temperatures below 62°.

t.°	Correction.	t.°	Correction.	t.°	Correction.	t.°	Correction.
32	—0·0019	48	—0·0012	64	+0·0002	80	+0·0023
33	·0019	49	·0011	65	·0003	81	·0024
34	·0018	50	·0011	66	·0004	82	·0026
35	·0018	51	—0·0010	67	·0005	83	·0027
36	·0018	52	·0009	68	+0·0007	84	·0029
37	·0017	53	·0008	69	·0008	85	+0·0031
38	—0·0017	54	·0008	70	·0009	86	·0032
39	·0017	55	·0007	71	·0010	87	·0034
40	·0016	56	—0·0006	72	·0012	88	·0036
41	·0016	57	·0005	73	·0013	89	·0037
42	·0015	58	·0004	74	+0·0014	90	+0·0039
43	·0015	59	·0003	75	·0016	91	·0041
44	—0·0014	60	·0002	76	·0017	92	·0042
45	·0014	61	—0·0001	77	·0018	93	·0044 ·
46	·0013	62	0·0000	78	·0020	94	·0046
47	—0·0013	63	+0·0001	79	+0·0021	95	+0·0048

158. NEWMAN'S SELF-REGISTERING TIDE-GAUGE.

At places where the phenomena of the tides are of much maritime importance, a continuous series of observations upon the rise and fall, and times of change, is essentially necessary as a basis for the construction of good tide tables; and as such observations should also be accompanied with the registration of atmospheric phenomena, we have no hesitation in inserting a description of an accurate self-registering tide-gauge.

The tide-gauge, as shown in the illustration, consists of a cylinder, A, which is made to revolve on its axis once in twenty-four hours by the action of the clock, B. A chain, to which is attached the float, D, passes over the wheel, C, and on the axis, of this wheel, C (in about the middle of it) is a small toothed wheel, placed so as to be in contact with a larger toothed wheel carrying a cylinder, E, over which passes another smaller chain. This chain, passing along the upper surface of the cylinder, A, and round a second cylinder, F, at its further end, is acted on by a spring so as to be kept in a constant state of tension. In the middle of this chain a small tube is fixed for carrying a pencil, which, being gently pressed down by means of a small weight on the top of it, performs the duty of marking on paper placed round the cylinder the progress of the rise or fall of the tide as the cylinder revolves, and as it is drawn by the chain forward or backward by the rise or fall of the float. The paper is prepared with lines equidistant from each other, to correspond with the hours of the clock, crossed by others showing the number of feet of rise and fall.

The cylinder while in action revolves from left to right to a spectator facing the clock, and the pencil is carried horizontally along the top of the cylinder; and the large wheel being made to revolve by the rise and fall of the float, turns the wheel

with the small cylinder, E, attached to it. If the tide is *falling*, the small chain is wound round the cylinder, E, and the pencil is drawn towards the large wheel; but if the tide is *rising*, the small chain is wound on to the cylinder, F, by means of the spring contained in it, which constantly keeps it in a state of tension. Thus, by means of the rise and fall of the tide, a lateral progress is given to the pencil, while the cylinder is made to revolve on its axis by the clock, so that a line is traced on the paper showing the exact state of the tide continuously, without further attention than is necessary to change the paper once every day, and to keep the pencil carefully pointed; or a metallic pencil may be used, which will require little, if any, attention.

A good self-registering tide-gauge is a valuable and important acquisition wherever tidal observations are required, and the only perfectly efficient instrument of this kind is that invented by the late Mr. John Newman, of Regent Street, London. It is now in action in several parts of the world, silently and *faithfully* performing its duty, requiring no other kind of attention than that of a few minutes daily, and thus admitting the employment of the person on any other service whose duty it would otherwise have been to have registered the tide. It has done much by its faithful records in contributing to the construction of good tide tables for many places; for those unavoidable defects dependent on merely watching the surface on a divided scale are set aside by it, all erroneous conclusions excluded, and a true delineation of Nature's own making is preserved by it for the theorist.

ADDENDA.

1. French barometers are graduated to millimetres. An English inch is equal to 25·39954 millimetres. Hence, 30 inches on the English barometer scales correspond to 762 millimetres on the French barometer scales. Conversions from one scale to another can be effected by the following formulæ:—

(1) Inches = millimetres divided by ... 25·39954
(2) Millimetres = inches multiplied by ... 25·39954

Of course, a table of equivalent values should be drawn up and employed, when a large number of observations are to be converted from one scale to the other.

2. In Germany, barometers are sometimes graduated with old French inches and lines,—the vernier generally indicating the tenth of a line.

OLD FRENCH LINEAL MEASURE.

					English Inches.
1 douzième, or point ...	...	...	...	=	0·0074
12 points ...	...	= 1 ligne	...	=	0·0888
12 lignes ...	...	= 1 pouce	...	=	1·065765
12 pouces ...	...	= 1 pied	...	=	12·7892
1 pied ...	...	= 324·7 millimetres.			

" The Germans indicate inches by putting two accents after the number; lines, by putting three accents; 27″ 3‴·85, means 27 inches 3 lines 85 hundredths of a line; more frequently, they give the height in lines, and the preceding number becomes 327‴·85."—*Kaemtz.*

3. *Rule for finding Diameter of Bore of a Barometer Tube.*

" If the maker has not taken care to measure the interior diameter directly, it may be deduced from the exterior diameter. The exterior diameter is first measured by calipers, and, by deducting from this diameter 0·1 of an inch for tubes from ·3 to ·5 of an inch in external diameter, we have an approximation to the interior diameter of the tube."—*Kaemtz.*

4.   WIND SCALES.

Sea Scale.		Wind.		Land Scale.		
0 to 3	=	Light	=	0	to	1
3 ,, 5	=	Moderate	=	1	,,	2
5 ,, 7	=	Fresh	=	2	,,	3
7 ,, 8	=	Strong	=	3	,,	4
8 ,, 10	=	Heavy	=	4	,,	5
10 ,, 12	=	Violent	=	5	,,	6

Pressure in Pounds (Avoirdupois)		(Land Scale).		Velocity in Miles (Hourly).
¼	=	1	=	10
5	=	2	=	32
10	=	3	=	45
21	=	4	=	65
26	=	5	=	72
32	=	6	=	80

5.   Letters to Denote the State of the Weather.

> *b* denotes blue sky, whether with clear or hazy atmosphere.
> *c* ,, cloudy, that is detached opening clouds.
> *d* ,, drizzling rain.
> *f* ,, fog.
> *h* ,, hail.
> *l* ,, lightning.
> *m* ,, misty, or hazy so as to interrupt the view.
> *o* ,, overcast, gloomy, dull.
> *p* ,, passing showers.
> *q* ,, squally.
> *r* ,, rain.
> *s* ,, snow.
> *t* ,, thunder.
> *u* ,, ugly, threatening appearance of sky.
> *v* ,, unusual visibility of distant objects.
> *w* ,, wet, that is dew.

A letter repeated denotes much, as *r r*, heavy rain; *f f*, dense fog; and a figure attached denotes duration in hours, as 14 *r*, 14 hours rain.

By the combination of these letters, all the ordinary phenomena of the weather may be recorded with certainty and brevity.

EXAMPLES.— *b c*, blue sky with less proportion of cloud.   2 *r r l l t*, heavy rain for two hours, with much lightning, and some thunder.

The above methods of recording the force of wind and state of weather were originally proposed by Admiral Sir Francis Beaufort. They are now in general use at sea, and by many observers on land.

6.   Table of Expansion by Heat from 32° to 212° F.

Platinum	...	...	...	0·0008842 of the length.
Glass, Flint	...	...	...	0·0008117 ,,
,, with Lead	...	...	0·0008622 ,,
Brass ...	...	...	...	0·0018708 ,,
Mercury	...	...	...	0·0180180 ,,
Water	...	...	...	0·0488200, from 39° to 212°
Alcohol	...	...	...	0·1100 ,, 32° to 174°
Nitric Acid	...	...	...	0·1100
Sulphuric Acid	...	...	0·0600

7.   Table of Specific Gravity of Bodies at 32° F. except water, which is taken at 39°·4.

Water	...	...	...	1·000
Alcohol, pure	...	...	0·791
,, proof	...	...	0·916
Mercury	...	...	...	13·596
Glass	...	...	...	8 to 2·7
Brass	...	...	...	7·8 to 8·54
Platinum	...	...	...	21 to 22·00

Weight of a cubic foot of water, at the temperature of comparison, 62·425 ℔s. avoirdupois.

The pound avoirdupois contains 7,000 grains.

Air is 813·67 times lighter than water.

The linear expansions are the mean values of the results of various experimentalists. The specific gravities are as given in Professor Rankine's *Applied Mechanics*.

8.   Important Temperatures.   Under the circumstances of—

Water	...	...	boiling at	...	...	212°
Mercury	...	...	boils at	...	...	660
Sulphuric Acid	...	,,	...	..	590
Oil of Turpentine	...	,,	..	...	560
Nitric Acid	...	,,	...	...	242
Alcohol	...	...	,,	...	...	174
A Saturated Solution of Salt	,,	...	...	218
Vital Heat	...	...	...	...	...	96
Olive Oil begins to solidify	...	...	...	36
Fresh Water freezes	...	...	...	32
Sea Water freezes	...	...	...	...	28
Mercury freezes	...	...	...	...	— 89

9. TABLE OF METEOROLOGICAL ELEMENTS, FORMING EXPONENTS OF THE CLIMATE OF LONDON.

1841 to 1861. MONTHS.	Mean Height of Barometer, reduced to 32° F., at the mean sea-level.	Mean Monthly Range of Barometer.	Mean of all the Highest Temperatures.	Mean of all the Lowest Temperatures.	Mean Temperature.	Mean Temperature of Dew-point.	Mean Degree of Humidity.	Mean Number of Rainy Days.	Average Rainfall.	Average Amount of Cloud (10=overcast).	Prevalent Winds.	Sun above the Horizon on Middle Day.	REMARKS.
	Inches.	Inches.	°	°	°	°			In.			Hours	
January	29·932	1·44	48·2	33·7	38·3	35·4	89	11	1·8	7·7	W. to N.	8¼	The majority of the nights are frosty.
February	29·962	1·22	44·7	33·2	38·4	34·4	85	10	1·6	7·4	S. to W.	10	10 frosty nights on the average.
March	29·967	1·23	50·0	35·3	41·7	36·4	82	10	1·5	6·6	N. to E.	12	12 ditto ditto. Strong winds.
April	29·907	1·06	56·8	38·6	46·3	39·9	79	11	1·8	6·1	N. to E.	14	6 ditto ditto.
May	29·931	1·02	64·4	44·2	52·8	45·5	76	11	2·1	6·1	S. to W.	15½	Very rarely frost.
June	29·960	0·89	71·2	50·2	59·2	50·8	74	11	1·9	6·1	W. to N.	16¼	Sun attains greatest North Declination, 21st.
July	29·970	0·79	73·8	53·2	61·9	53·9	76	11	2·7	6·9	W. to N.	16	
August	29·954	0·97	72·8	53·4	61·3	54·1	77	12	2·4	6·5	W. to N.	14¼	
September	29·997	0·95	67·4	48·9	56·9	51·1	81	13	2·4	5·9	S. to W.	12¼	A few frosty nights. Heavy gales.
October	29·860	1·38	58·3	43·7	50·2	46·0	87	12	2·8	6·9	S. to W.	10¼	11 nights frosty.
November	29·929	1·53	49·3	37·7	43·4	40·1	89	12	2·4	7·2	S.W.	9	Sun attains greatest South Declination, 21st.
December	29·979	1·52	45·0	35·5	40·1	36·9	89	12	1·9	7·4	W.	8	
Year	29·946	1·16	58·0	42·3	49·2	43·7	82	133	25·8	6·7	...	...	
	1	2	3	4	5	6	7	8	9	10	11	12	13

In the above Table, columns 1 to 10 are results obtained at the Royal Observatory, Greenwich, by J. Glaisher, Esq., F.R.S. The data contained in columns 2 and 10, are deduced from observations extending over the years 1841 to 1855 inclusive, and are copied from Edward Hughes' *Third Reading Book;* the other columns are results of observations made during the twenty years ending 1861. The rest of the information is from Luke Howard's *Climate of London.*

These valuable data indicate the characteristics of the weather in each month in the suburbs of London, and will be found tolerably accurate as indications of weather, and serviceable as standards for comparisons of observed results, at most places in England.

STANDARD WORKS ON METEOROLOGY

SUPPLIED BY

NEGRETTI & ZAMBRA.

THE WEATHER BOOK:
A MANUAL OF PRACTICAL METEOROLOGY.
By Vice-Admiral FitzRoy, F.R.S., M.I.F., &c.
Price, £0 15 6

THE LAW OF STORMS,
By H. W. Dove, F.R.S.
Translated by R. H. Scott, M.A.
Price, £0 10 6

L. F. KÆMTZ'S "COMPLETE COURSE OF METEOROLOGY,"
Translated by C. V. Walker, Esq.
Price, £0 12 6

PRACTICAL METEOROLOGY,
By John Drew, Ph.D., F.R.A.S.
Price, £0 5 0

HYGROMETRICAL TABLES,
Adapted to the use of the Wet and Dry Bulb Thermometer,
By James Glaisher, Esq., F.R.S.
Price, £0 2 6

TABLES OF THE CORRECTIONS FOR TEMPERATURES,
To reduce observations to the 32° Fahrenheit, for Barometers with brass scales
extending from the cistern to the top of the mercurial column,
By James Glaisher, Esq., F.R.S.
Price, £0 1 0

TABLE OF THE DIURNAL RANGE OF THE BAROMETER,
By James Glaisher, Esq., F.R.S.

Price, £0 0 6

TABLES FOR CALCULATION OF HEIGHTS FROM OBSERVATIONS
ON THE BOILING-POINT OF WATER,
Adapted to the use of Negretti and Zambra's Boiling-point Apparatus.

Price, £0 1 0

A THERMOMETRICAL TABLE,
ON THE SCALES OF FAHRENHEIT, REAUMUR, AND CENTIGRADE,
By Alfred S. Taylor, Esq., M.D., &c.

Price, in Sheet, with explanatory Pamphlet, £0 1 6

METEOROLOGICAL TABLES,
For the reduction of Barometrical and Hygrometrical Observations, Determination of Heights by the Barometer and Boiling-point Thermometer, &c.

By G. Harvey Simmonds, M.B.M.S.

Price, £0 2 6

BAROMETER MANUAL,
Compiled by Vice-Admiral FitzRoy, F.R.S.,

For the Board of Trade.

Price, £0 0 6

POCKET METEOROLOGICAL REGISTER AND NOTE-BOOK,
With Diagrams for exhibiting the Fluctuations of Barometer, &c.

Printed on metallic paper. *Price,* with Pencil, £0 3 0

LONDON:

PRINTED BY STRAHAN AND WILLIAMS,

7 LAWRENCE LANE, CHEAPSIDE, E.C.

NEGRETTI & ZAMBRA'S
PRICE LIST

OF

BAROMETERS, THERMOMETERS,

AND

HYGROMETERS,

RAIN, VACUUM, AND STEAM PRESSURE GAUGES, &c.

Separate from Standard Meteorological Instruments, there are others which partake of a more general character, and are used by the husbandman, the mariner, and the artizan, entirely as "WEATHER GLASSES." Each Instrument is arranged to suit the particular work it has to perform, and may be depended upon for accuracy.

Barometers can be supplied to order of any style of architecture, so as to correspond with the furniture of libraries, halls, &c. Barometers are now mounted in so many varied styles, both plain and carved, that the following are given only as being those most in demand.

BAROMETERS.

DIAL BAROMETERS OR WEATHER GLASSES.

			£	s.	d.
1	8-inch **Silvered-brass Dial Barometer**, mounted in common mahogany or rose-wood frame, with or without mirror, hygrometer, and level . . 25/ to		1	10	0
2	8-inch ditto, square bottom . . . . 30/ to		2	2	0
3	8-inch ditto, superior finish, best engraving, and large tube . .. 50/ to		3	10	0
4	10-inch **Silvered brass Dial Barometer**, in common mahogany or rosewood frame, with or without mirror, hygrometer, and level . . . 35/ to		2	10	0
5	10-inch ditto, square bottom. . . . . 50/ to		3	3	0
6	10-inch **extra best** ditto, both as regards frame, dial, engraving, and tube 70/		5	5	0
7	8-inch **Scroll Pattern Dial Barometer**, best mahogany or rosewood frame, silvered brass dial, with thermometer. . . . . 50/ to		3	10	0
8	8-inch ditto, with figures and divisions on porcelain dial, and thermometer in front		3	15	0
9	8-inch ditto, rosewood frame inlaid with *pearl*, with silvered brass dial . 60/ to		4	0	0
10	10-inch ditto, best rosewood frame, inlaid with *pearl or brass*, the dial of silvered brass, with thermometer, double basil ring, and polished-edge plate glass, superior engraving and finish		6	6	0
11	10-inch **Dial Barometer**, in solid oak, mahogany, or walnut frame, either plain or chastely carved, of the very best construction . . . . £5 5s. to		8	8	0
	12-inch dial ditto at proportionate prices.				
12	14-inch **Dial Barometer**, best rosewood frame, inlaid with *pearl or brass*, and an eight-day pendulum clock fitted in the frame, thermometer, &c. . £15 to		20	0	0
	Suitable for club-houses, mansions, &c.				

N.B.—DIAL BAROMETERS required for transmission to distant parts, such as India or the Colonies, should be ordered *expressly*, as in that case they will be furnished with a steel stopcock, to render them portable more effectually than can be done by the old method of plugging the tube. These additions will enhance the price of each barometer, 7s. 6d.

PORTABLE PEDIMENT BAROMETERS.

		£	s.	d.
13	Small Model Barometer, with ivory scale and sliding index	0	10	6
14	Ditto, larger size, with vernier and thermometer, also screw to render it portable	1	1	0
15	Pediment Barometers, with larger tube, ivory scale, thermometer, *sliding vernier* reading to $\frac{1}{100}$ of an inch, the tube visible throughout the whole length, and portable screw . . . . 42/ to	2	10	0
16	Ditto ditto, with glass cover over the face, rackwork vernier and thermometer .	2	15	0
17	Portable Pediment Barometer, in oak, mahogany, or rosewood frame, the tube of large diameter, covered entirely, with thermometer on front . . .	3	0	0
18	Ditto ditto, with square moulded top, larger tube, and two verniers . 70/ to	4	0	0
19	Ditto ditto, bow front, with 0·40-inch diameter tube, double rackwork vernier, the scales elegantly engraved on ivory plates, and thermometer in oak, mahogany, walnut, or rosewood frame . . . . 90/ to	5	0	0
20	Portable Pediment Rosewood Barometer, elegantly inlaid with pearl, thermometer in front, ivory scale, rackwork vernier . . . . £4 to	5	5	0
21	Portable Pediment Barometers, in solid oak, rosewood, or mahogany, elegantly carved, of various designs, fitted up in the very best manner . £5 5s. to	12	12	0
22	Large Barometers, fitted in oak, walnut, or ebony frames, the tube 0·5-inch in internal diameter, and the cistern presenting a large area, to ensure uniformity in reading, ivory or *patent porcelain* scales, and two verniers . £6 6s. to	8	8	0

22* **The FISHERMAN'S and LIFE BOAT STATION BAROMETER,** as made by Negretti and Zambra especially for the Board of Trade and Royal Life Boat Institution, to be fixed at all the principal sea ports, fishing and life boat stations 5 5 0

This barometer consists of a tube with very large bore, and an accurate thermometer, mounted in a solid oak frame, firmly screwed together, with scales engraved on *porcelain* by Negretti and Zambra's new patent process, the vernier reading to 100-ths of an inch. It is strongly recommended as a good sound working instrument, admirably adapted for use in public institutions.

23 **Negretti and Zambra's Farmers' Barometer,** for ascertaining the humidity of the atmosphere, the general character of the weather, and the approach of Wind or Rain. The Farmers' Barometer combines three distinct instruments—the Barometer, the Thermometer, and the Hygrometer, and is equally valuable to the Agriculturist and the Invalid 2 10 0

STANDARD AND MOUNTAIN BAROMETERS.

		£	s.	d.
23*	Negretti and Zambra's Standard Barometer, as used and recommended by the British Meteorological Society . . . . . . £8 8s. to	10	10	0
23**	Large ditto ditto for Observatory use . .	20	0	0
24	Negretti and Zambra's Standard Mountain Barometer, the only portable instrument made possessing all the essentials of a *standard barometer* . .	8	8	0

With mahogany tripod stand, which forms a travelling case.

25 Gay-Lussac's Syphon Barometer, with his original improvement for excluding the air from the column of mercury; rackwork verniers at each extremity; a thermometer in the centre; the tube so constructed that the barometer will travel safely; reading to $\frac{1}{100}$th of an inch, either from the centre by adding the two readings together, or from the whole length of the column by substracting the bottom reading from the top 7 7 0

In brass frame and leather sling case.

26 Ditto ditto, with brass tripod stand and superior finish . . . 10 10 0

27 Mountain Barometer, with turned wooden frame, having brass shield and portable screw; brass scale, reading by vernier to $\frac{1}{500}$th of an inch; and detached thermometer 4 4 0

MARINE BAROMETERS.

		£	s.	d.
28	Marine Barometer, in mahogany, rosewood, or oak frame, with ivory scale, sliding vernier, thermometer, and brass arm gimble, for suspension, complete, 42/ to	2	10	0
29	Marine Barometer, round, moulded, or carved top, with rackwork vernier, thermometer in the front and capillary tube to prevent the ingress of air into the column, even during the most violent oscillations of a storm	3	3	0

		£	s.	d.
30	Ditto ditto, in carved frame, or inlaid with pearl . . . 70/ to	4	4	0
31	Marine Barometer, with Sympiesometer in front, combining the mercurial with the air barometer, and serving to check one with the other . . £5 5s. to	6	6	0
32	Board of Trade Standard Marine Barometer, as made by Negretti and Zambra for Her Majesty's Government	4	4	0
32*	Admiral Fitzroy's Marine Gun Barometer	5	5	0
	In travelling case with lock and key.			

ANEROID BAROMETERS.

		£	s.	d.
33	Aneroid Barometers, with card dials	2	10	0
34	Ditto, with silvered metal dials	3	0	0
35	Ditto, with metal dials and thermometer	3	10	0
35*	Ditto, for measuring heights, with compared and corrected scale, as supplied by Negretti and Zambra to the Royal Navy	4	10	0
36	Negretti and Zambra's Pocket Aneroid Barometer, especially convenient for Travellers	3	0	0
36*	Ditto ditto ditto smallest size, and carefully corrected and adjusted for determining altitudes	5	5	0
37	Bourdon's Metallic Barometers, with plate glass front . . . £4 to	4	4	0
38	Table Stands, for Aneroid or Bourdon's barometers . . . 12/6 to	1	10	0

SYMPIESOMETERS AND STORM GLASSES.

		£	s.	d.
39	Sympiesometers, in rosewood or mahogany frames, with registering index and plate glass front . . . . 50/ to	3	3	0
40	Ditto ditto, with rackwork movement, large size and best make . . .	4	10	0
41	Small Sympiesometer, suitable for travelling, and taking altitudes . . .	3	15	0
42	Storm Glass, or Chemical Barometer	0	5	6
43	Ditto ditto, mounted in mahogany window bracket	0	10	6
44	Ditto ditto, mounted with a thermometer on a boxwood scale . . 6/6 to	0	10	6
45	Ditto ditto ditto mounted in window bracket . .	0	15	0

THERMOMETERS.

		£	s.	d.
46	6 or 8-inch Boxwood Scale Thermometer . . . . . 1/ to	0	1	6
47	6 or 8-inch ditto, with French polished scale . . . 1/6 to	0	2	6
48	6 or 8-inch ditto, with enamel tube, best make	0	3	6
49	6 or 8-inch ditto, with enamel tube, the scale bevelled at the edges, very superior .	0	4	6

PORTABLE OR POCKET THERMOMETERS.

		£	s.	d.
50	3 or 4-inch Ivory or Metal Scale, in morocco leather case . . . 3/6 to	0	5	6
51	6-inch ditto ditto . . . . 4/6 to	0	7	6
52	8-inch ditto ditto . . . . 7/6 to	0	10	6
53	Oval Boxwood Thermometer, with tube and bulb sunk in the solid wood, to prevent breakage in travelling	0	7	6
54	Oval Ivory ditto ditto	0	15	0
55	Circular Thermometer, with ivory scale, in leather case, 2-inches in diameter .	0	10	6
56	Ditto, 3-inches in diameter, with compass in centre . . . 14/ to	0	18	0
57	Ditto, 3-inches in diameter, with compass and sun dial in centre . . 15/ to	1	10	0
58	Compound Bar Metallic Thermometer, in the form of a watch . 50/ to	3	0	0
59	Negretti and Zambra's improved Travelling Thermometer, in plated metal or silver case . . . . 10/6 to	1	0	0
	Not larger than a pencil case; accurately divided on its own stem.			
60	Negretti and Zambra's ditto ditto, in German silver revolving case . 10/6 to	0	15	0

DRAWING ROOM THERMOMETERS.

			£	s.	d.
61	6 or 8-inch Thermometer, Ivory Scale on Ebony Back .	6/6 to	0	10	6
62	10-inch ditto, extra large	. .	0	15	0

MANTEL OR TABLE THERMOMETERS.

63	Ivory Scale on Ebony Stand, with glass shade	7/6 to	0	12	6
64	Ditto, on solid ivory stand	18/ to	1	10	0
65	Ditto, handsomely engine-turned, and ornamented in numerous designs .	30/ to	3	0	0
66	Ditto, with compass or sun dial at top	25/ to	2	10	0
67	10 or 12-in. Glass Scale Thermometers, superior workmanship and elegant appearance, on oak, mahogany, or ebony backs, with Negretti and Zambra's enamelled tubes, suited for halls or passages . . . . 25/, 30/ to		2	2	0

WINDOW THERMOMETERS FOR OUT-DOOR USE.

68	Window Thermometer, 8-inch Ivory Scale, enclosed in glass cylinder, on oak bracket, with metal top	.	0	10	6
69	10-inch ditto ditto	15/ to	0	18	0
70	12-inch ditto ditto	18/ to	1	0	0
71	10-inch, 12-inch, or 14-inch Patent Porcelain Scale mounted on oak or mahogany bracket, with brass supports for fixing at any angle . .	15/ to	1	5	0
72	Ditto Glass Scale, with enamel tube, the scale divided by engine, and handsomely mounted on oak bracket, with brass supports	30/ to	2	10	0

SELF-REGISTERING THERMOMETERS FOR HEAT.

73	Rutherford's Maximum Thermometer, on boxwood or metal scale, with steel index	3/6 to	0	7	6
74	Ditto, on Negretti and Zambra's patent porcelain scale, in metal case .	.	0	10	6
75	Phillip's Maximum Thermometer on boxwood or metal scale, with air index .		0	7	6
76	Ditto, on Negretti and Zambra's patent porcelain scale, in metal case .	. .	0	10	6
77	Negretti and Zambra's Patent Maximum Thermometer*	.	1	1	0
78	Ditto ditto, for solar radiation	.	1	5	0
79	Ditto ditto, in vacuum	.	1	10	0

    * This Instrument is the only Maximum Thermometer that can be recommended, as unless it be broken it cannot be put out of adjustment.

SELF-REGISTERING THERMOMETERS FOR COLD.

80	Rutherford's Minimum Thermometer, on boxwood or metal scale .	. 3/6,	0	7	6
81	Ditto, on Negretti and Zambra's patent porcelain scale, in metal case	. .	0	10	6
82	Negretti and Zambra's Standard Minimum Thermometer . . .	.	1	1	0
83	Ditto ditto, for terrestrial radiation	.	1	5	0
84	Negretti and Zambra's Patent Mercurial Minimum Thermometer . .	.	2	2	0
85	Negretti and Zambra's Horticultural Self-registering Minimum Thermometer. The scale is made of stout zinc, and enclosing the tube, perfectly protecting it from injury, the figures and divisions are raised for easy reading the indications		0	3	6

    Strongly recommended in all the leading Horticultural Journals as the cheapest and best registering thermometer of the kind for garden purposes.

SELF-REGISTERING THERMOMETERS FOR HEAT AND COLD.

		£	s.	d.
86	Six's Registering Thermometer, 10-inch on Boxwood or Metal Scale, in japanned case . . . . 14/ to	0	15	0
87	Ditto 12 or 14-inch ditto ditto	1	1	0
88	Ditto 10-inch on Ivory ditto	1	1	0

Copper cases 3s. extra.

		£	s.	d.
89	Ditto, Ivory scale on Ebony back, for indoor use . . . . 25/ to	1	10	0
90	Ditto, with Negretti and Zambra's patent porcelain scale, on oak back, for window . . . . 15/ to	1	5	0
91	Ditto ditto, in metal cases . . . . . . 21/ to	1	10	0
92	Negretti and Zambra's small size Patent Maximum and Minimum Thermometer, arranged in a mahogany case, suited for travellers to whom size and weight is an object . . . . 35/ to	2	2	0
92*	Ditto ditto larger size	2	10	0

BOTANICAL THERMOMETERS.

		£	s.	d.
93	8-in. Boxwood Scale in japanned metal cases, range of scale 0 to 120° or 150°	0	3	6
94	10-in. ditto ditto	0	5	6
95	12 to 14-in. Boxwood Thermometers in japanned cases . . 7/6 to	0	10	6
96	8-in. on Negretti and Zambra's patent porcelain scales. . . .	0	4	6
97	10-in. ditto ditto	0	5	6
98	12-in. ditto ditto, and 14-in. ditto ditto	0	7	6
99	Hot Bed Thermometer, in metal mounting	0	12	0
99*	Ditto ditto in mahogany frame, encased in brass cylinder 25/ to	1	10	0
100	Thermometer, for ascertaining the temperature of the earth at various depths, 5-ft. long . . . . 25/ to	1	10	0
101	Delicate Thermometers, for inserting in the stems and flowers of growing plants	0	10	6

BREWERS' THERMOMETERS.

		£	s.	d.
102	8-in. Brewers' Thermometer, brass Scales, in japanned metal case 3/ to	0	5	0
103	10-in. ditto ditto . . . . 3/6 to	0	6	0
104	12 and 14-in. ditto . . . . 5/6 to	0	7	6
105	8-in. ditto, enamelled tubes, in copper cases . . . . 4/6 to	0	6	6
106	10-in. ditto ditto . . . . 5/6 to	0	7	6
107	12 and 14-in. ditto ditto . . . . 7/6 to	0	10	6
108	Brewers' Standard Reference Thermometers	1	1	0
109	8-inch Brewers' Thermometer, Porcelain Scales, Negretti and Zambra's patent, range of scale 212°, in japanned metal cases	0	4	6
110	10-inch ditto ditto	0	5	6
111	12 and 14-in. ditto	0	7	6
112	8-in. ditto, in copper cases	0	5	6
113	10-in. ditto ditto	0	7	0
114	12 and 14-in. ditto	0	10	6
115	Gyle Tun Thermometers, according to length . . . . 15/ to	1	10	0

BATH THERMOMETERS,

SAME PRICE AND FORM AS BREWERS' THERMOMETERS.

		£	s.	d.
116	Floating Thermometers, for keeping constantly in water . .	0	7	6

CHEMICAL AND SURGICAL THERMOMETERS.

		£	s.	d.
117	Chemical Thermometer, with Boxwood Scale, graduated to 300° or 400°, the bulb projecting below the scale	0	4	6
119	Ditto, with brass hinge jointed boxwood scale, to 300°	0	7	6
120	Ditto, superior enamel tube, and French polished, 600° . . . 10/6 to	0	12	6
121	Chemical Thermometer, graduated on stem for inserting in the tubulure of retorts, to 400°	0	5	6
122	Ditto ditto, to 600°	0	7	6
123	Ditto, best make and engine divided	0	10	6
124	Thermometers insulated in Glass Cylinders, for acids			
125	Ditto, 40° to 300°	0	5	6
126	Ditto, 40° to 600°	0	7	6
127	Delicate Thermometers of various forms, for ascertaining the heat of the body . . . . 10/6 to	0	15	0
128	Ditto, Negretti and Zambra's self-registering	1	1	0

MARINE THERMOMETERS.

		£	s.	d.
129	Deep Sea Sounding Thermometer, in copper case	1	10	0
130	Ditto, self-registering, with valves and Negretti and Zambra's *patent porcelain scales*, as supplied to the Board of Trade and Admiralty	2	10	0
131	Board of Trade Thermometer, the scale divided on its own stem, and fixed on Negretti and Zambra's *patent porcelain scales*, in japanned metal case . .	0	7	6
132	Ditto ditto in copper case	0	8	6

THERMOMETERS FOR SPECIAL PURPOSES.

		£	s.	d.
133	Oven Thermometers for high temperatures, on stand	0	12	6
134	Ditto, Negretti and Zambra's patent self-registering . . . . 21/ to	1	10	0
135	Dairy Thermometers, with ivory mountings . . . . . 7/6 to	0	10	6
136	Beehive Thermometers	0	5	6
137	Soap Boilers' Patent Thermometers	0	7	6
138	Dentists' Thermometers for vulcanizing process . . . 7/6 to	1	5	0
139	Sugar-Boiling Thermometer, 3 to 4-ft. long, graduated to 280° . 18/ to	1	10	0
140	Steam Pressure Thermometers, in strong brass cases, with plug for closing the boiler when the thermometer is not in use . . . . . 25/ to	2	0	0
141	Vacuum Pan Thermometers, in strong brass cases, with hinged doors 30/ to	2	10	0
142	Thermometer, in brass frame, for hot air apparatus and high temperatures .	1	10	0
143	Super-heated Steam Thermometers, with patent porcelain scales, iron or brass mountings . . . . 25/ to	1	10	0
144	Standard Thermometers	2	2	0
145	Boiling Point Thermometers, for determining heights by observing the boiling point of water, divided to 0·1 degrees	1	10	0
146	Leslie's Differential Thermometers, for experiments on radiation 15/ to	1	0	0
147	Air Thermometers, fitted on scale for delicate experiments . . 7/6 to	1	1	0

PYROMETERS.

		£	s.	d.
148	Daniell's Pyrometer, for indicating high temperatures	4	10	0
149	Wedgwood's ditto ditto			
150	Ferguson's, for showing the difference of expansion in metals . . .	3	10	0

HYGROMETERS.

		£	s.	d.
151	The Oat-Beard Hygrometer, or Damp Detector, for ascertaining the comparative dryness and dampness of different apartments, beds, etc.	0	10	6
	Size of a small watch, for the pocket.			
152	Wet and Dry Bulb Hygrometer, sometimes called Mason's Hygrometer	1	1	0
153	Do. arranged on brass tripod stand, with folding legs, and metal cover for protection	3	3	0
154	Standard Wet and Dry Bulb Hygrometer	2	2	0
155	Daniell's Hygrometer, for ascertaining the dew point by direct observation	2	12	0
156	Regnault's Condenser Hygrometer	3	3	0
156*	Portable Wet and Dry Bulb Hygrometer, in mahogany case	2	2	0
	This is a companion instrument to No. 92.			

STEAM, VACUUM, HYDRAULIC, AND RAIN GAUGES.

STEAM GAUGES.

		£	s.	d.
157	Steam Gauges, from 15 to 120-lbs., with mercurial tube, and union joint, adapted either at bottom or at the side of the frame; in mahogany frame	2	2	0
158	Ditto, ditto, in iron ditto	2	7	0
159	Ditto, ditto, in brass ditto	2	10	6
160	Thermometric Pressure Gauge, for showing the pressure of vapour by taking its temperature . . . 25/ to	2	0	0
161	Schaeffer's Patent Metallic Pressure Gauge, with graduated dial, ranging from zero to 30, 50, 80, 100, and upwards, to 300 pounds on the square inch	3	10	0
162	Ditto, for hydraulic pressure, to 3,000 pounds	10	10	0
163	Bourdon's Pressure Gauge, with metal taps, adapted for all pressure below nine atmospheres, in varnished cast iron box	3	5	0
164	Ditto ditto, in round brass box	4	8	0
165	Small portable ditto, to ten atmospheres, with steel screws and plates for fixing to the machines, in case complete	3	5	0
166	Ditto ditto, to eighteen atmospheres	3	10	0
166*	Steam Engine Indicator . . . £5 5s. to	6	6	0

VACUUM GAUGES.

		£	s.	d.
167	Vacuum Gauges in mahogany or oak frame, for general purposes	2	2	0
168	Ditto ditto, in iron frames	2	7	6
169	Ditto ditto, in brass ditto	1	10	6
	These Gauges are the same form as No. 157.			
170	Sugar Pan Vacuum Guage, the tube and scale enclosed in stout glass cylinder and brass frame, with door and hinges complete, with ground plug and stop-cock for fitting it to the pan . . . 30/ to	2	10	0

RAIN GAUGES.

		£	s.	d.
171	Glaisher's Rain Gauge, of japanned metal, with graduated measure	1	1	0
172	Ditto ditto of japanned copper	1	10	0
	Receiving Pots for ditto, extra.			

The graduated measure is divided into hundreths of an inch, according to the calculated weight of water, as determined by the area of the receiving surface.

£ s. d.

173 Rain Gauge, having a receiving surface of 12 inches diameter, and graduated glass gauge tube, or boxwood rule, divided to inches, tenths, and hundredths of an inch, showing by simple inspection the amount of rain fallen. In japanned metal with tap for emptying the gauge, with boxwood rule, 50/-; divided with glass tube 3 3 0

173* Ditto, in copper, extra 15/-

174 Howard's Portable Rain Gauge, with graduated glass measure . . . 0 15 0

175 Rain Gauge, simple form, fitted with divided gauge rod to show amount of rain fall in hundredths of an inch, and thousands of gallons per acre . . . 1 10 0

HYDROMETERS AND OTHER INSTRUMENTS FOR ASCERTAINING THE SPECIFIC GRAVITY OF FLUIDS AND SOLIDS.

176 Hydrometers, specific gravity, for fluids heavier or lighter than water . . 0 6 6

177 Hydrometers for Spirits (Sykes's glass) 0 5 6

178 Ditto, with thermometer, in mahogany box 0 15 0

179 Sykes's Hydrometer (brass), as used by the Excise, complete in box, with tables and test jar 4 0 0

180 Brewer's Saccharometer, glass 0 5 6

181 Ditto, with thermometer in mahogany box 0 15 0

182 Ditto, brass, with thermometer, density rule, assay jar, and one weight, in mahogany box 4 10 0

183 Lactometer, for ascertaining the specific gravity of milk 0 5 0

184 Beaume's, Twaddle's, and other Hydrometers . . . . .

185 Specific Gravity Beads, for achohol, &c., in sets . . . 6/6 to 0 10 6

Just Published,

New and Revised Editions, Illustrated with Wood Engravings,

HOW TO FORETELL WEATHER:

A MANUAL COMPILED BY ADMIRAL FITZROY, F.R.S.

Price (Post-free) Sixpence.

THE MAGIC LANTERN, DISSOLVING VIEWS, AND OXY-HYDROGEN MICROSCOPE DESCRIBED,

Also, Directions for their Use with OIL LAMPS, OXY-CALCIUM and OXY-HYDROGEN LIGHT,

AND INSTRUCTIONS FOR PAINTING ON GLASS.

Price (Post-free) Sixpence.

NEW DESCRIPTIVE CATALOGUE,

Illustrated by upwards of Five Hundred Engravings,

IN WHICH WILL BE FOUND DETAILED PARTICULARS AS TO PRICE, SIZE, ETC., OF ALL

OPTICAL, MATHEMATICAL, AND PHILOSOPHICAL INSTRUMENTS.

Price (Post-free) Two Shillings and Sixpence.

NEGRETTI AND ZAMBRA,

OPTICIANS, AND METEOROLOGICAL INSTRUMENT MAKERS TO THE QUEEN,

Her Majesty's Government,

THE ROYAL OBSERVATORY, GREENWICH, &c., &c.

Chief Establishment and Manufactory—1, HATTON GARDEN, E.C.

Branch Establishments—59, CORNHILL, E.C., and 122, REGENT STREET, W.,

LONDON.

ALFRED BOOT, PRINTER, DOCKHEAD, SOUTHWARK, S.E.

www.ingramcontent.com/pod-product-compliance
Lightning Source LLC
Chambersburg PA
CBHW021807190326
41518CB00007B/486